The Tractor Ploughing Manual

The Tractor Ploughing Manual

Edited by Brian Bell MBE

Old Pond
PUBLISHING

ISBN 978-1-913618-11-7

A catalogue record for this book is available from the British Library

 Fox Chapel Publishing
903 Square Street
Mount Joy, PA 17552

Fox Chapel Publishers International Ltd.
20-22 Wenlock Rd.
London N1 7GU, U.K.

www.oldpond.com

We are always looking for talented authors. To submit an idea, please send a
brief inquiry to acquisitions@foxchapelpublishing.com.

Printed in the USA

Endpaper illustrations: Ransomes, Sims and Jefferies catalogue, 1886

Front cover: Ploughing match competitors in the reversible, mounted and
trailed conventional plough classes. Photographs courtesy of the Society of
Ploughmen.

Cover design and book layout by Liz Whatling

Contents

*The Cairn of Peace at the site of the 63rd World Ploughing Contest
held at Crockey Hill near York in 2016. The Cairn contains an engraved
stone from every country taking part in the contest.*

Introduction

Ploughing has been practised since Biblical days when man first used his own muscle power to scratch the soil with a suitably shaped piece of wood. Early Britons were required by law to fashion their own ploughs before being allowed to use these mainly wooden implements to till the land. Ploughs were made from iron by the late eighteenth century and chilled cast-iron ploughshares were already turning the soil when Nelson defeated the French navy at Trafalgar. Originally a village affair, ploughing matches have been held for over 150 years with the best ploughman enjoying great esteem locally. J. Allen Ransome wrote in 1843 that the object of these matches was to promote the art of ploughing and it has remained so to the present day. Many countries hold their own national championships with the winners taking part in the annual World Ploughing Match. First held in 1953 this event is hosted by a different country each year. The first British National Ploughing Match, organised by the British Ploughing Association, was held in Yorkshire in 1951. This annual event, arranged by The Society of Ploughmen since 1973, is still keenly contested today.

The Tractor Ploughing Manual has been written for the benefit of both the novice and the more experienced ploughing match competitor in the hope that it will help to promote the art of ploughing. An introduction to ploughs and the basic principles of commercial ploughing is followed by a survey of the types of tractor plough used at ploughing matches up and down the country. The main body of this book sets out to explain the

The 2016 World Ploughing Match was held at Crockey Hill, near York.

*A Society of Ploughmen practical seminar dealing with the finer points
of hydraulic, trailed and reversible match ploughing techniques.*

complexities of the various classes of competition ploughing. Further chapters cover a brief history of the plough, the Society of Ploughmen's rules for ploughing matches and a glossary of ploughing terms.

Preparation for the first edition of *The Tractor Ploughing Manual* was a team effort by Ken Chappell MBE, David Chappell and Peter Alderslade, who provided a great deal of technical information and ploughed the demonstration plots which were photographed by John Allen. Additional illustrations were provided by the Society of Ploughmen, Alan Jones and Roger Smith. Ken Chappell and Sue Frith have provided new photographs and up to date information for the second edition of *The Tractor Ploughing Manual*. New material in this edition includes the current rules for match ploughing, judges' scoring system and a list of ploughing organisations in many countries throughout the world.

Brian Bell, September 2020

I. Types of Plough

Ploughs have evolved over the years from horse-drawn and steerable tractor ploughs to trailed and later to the semi-mounted and fully-mounted conventional and reversible ploughs of today. The soil-working components and their basic settings are common to all types of mouldboard plough but the methods of adjustment vary according to type, make and model.

shaft attached at right angles to the plough beams provides the hitch points used to attach the plough to the lower left and right hydraulic lift arms and the top link is connected to the upper part of the headstock. Depending on the make and the type of tractor used with a mounted plough it may have one, two or three wheels but sometimes none at all.

MOUNTED CONVENTIONAL PLOUGHS

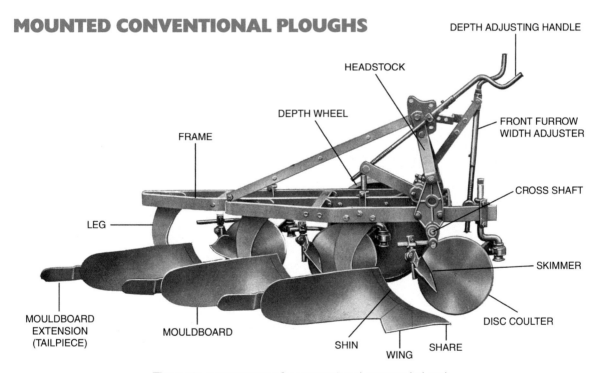

The main components of a conventional mounted plough.

The conventional mounted plough is attached to the tractor hydraulic three-point linkage. The plough frame consists of a separate beam for each plough body, the three-point linkage headstock and a curved or straight leg connecting each right-hand plough body to the beam. Ploughs may have one or more one-piece beams and legs or individual legs bolted to the beams. The cross

The Share
The share, or point, cuts the bottom of the furrow slice.

The Mouldboard
The mouldboard lifts and turns the furrow slice. The shape, size and appearance of the furrow slice depend on the profile of the mouldboard.

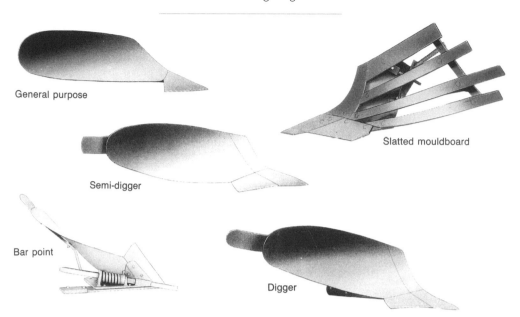

General purpose

Slatted mouldboard

Semi-digger

Bar point

Digger

The various types of plough body in current use have evolved from the traditional, general purpose semi-digger and digger plough bodies. Slatted mouldboards tend to break up the furrow slice more efficiently than a full mouldboard and improve soil movement across the mouldboard when ploughing sticky soils. The bar point body often has a spring-loaded ba which is moved forward as the point wears away. This body is ideal for soils with large stones near the surface.

The Landside

The landside absorbs the side thrust produced by the mouldboard. Some ploughs have a short, fixed landside and a hinged, spring-loaded rear furrow wheel or rolling landside. The vertical movement of a rolling landside provides a more rapid entry into work.

The Heel Iron

Bolted to the end of the rear landside, the heel iron supports some of the weight at the back of the plough.

The Tailpiece

The tailpiece, or mouldboard extension, helps to press down the furrow slice, especially when ploughing up grassland and heavy soils.

The Disc Coulter

The disc coulter cuts the side of the furrow slice about to be turned by the mouldboard. Some ploughs have a fixed knife coulter.

The Skimmer

The skimmer, or skim coulter, turns a small slice of the corner of the furrow about to be turned and throws it into the furrow bottom. This reduces the likelihood of weeds growing up between adjacent furrow slices.

The rear landside and hinged rear furrow or rolling landside of a Ferguson mounted plough.

Controls and Adjustments

Hitching

Hitching a mounted conventional plough to the three-point linkage is best done by first attaching the lower left lift arm, then the right lower lift arm and finally the top link. By using this method the right-hand lift rod levelling box can be used to align the right-hand lift arm with the cross shaft pin. The use of an adjustable top link makes it equally simple to align the top link pinhole with the holes in the head stock.

Disc Coulters

For normal work the bottom edge of the disc should be set 12 mm (½ in) to the unploughed side of the share. A stepped furrow wall indicates the disc is set too far towards the unploughed land and a ragged furrow wall means that the disc is too far towards the ploughed land. The height of the disc above the share will depend on ploughing depth. It will need to be about 12 mm (½ in) above the share when ploughing at a depth of approximately 15 cm (6 in) but it will need to be higher for deeper work. In all cases the disc should be set high enough to prevent the disc bearing from dragging along

on the previous furrow. This will cause undue wear on the housing and will lose points in a ploughing match. A knife coulter should be set with the leading tip of the coulter in the same position as the lowest edge of a disc coulter.

The disc coulter should normally run in a vertical position but a small amount of undercutting with the top edge slightly tilted towards the ploughed land can be an advantage when ploughing grassland.

For normal ploughing the disc coulter should be about 12 mm (½ in) above the share but it will need to be higher for deeper work.

Skim Coulter

The point of the skimmer share should be below and behind the disc coulter hub and should be set to work at no more than one third of ploughing depth. For example, when ploughing at 10 cm (4 in) the skimmer point should work at a depth of no more than 6 cm (2½ in). The skimmer only needs to be set low enough to prevent any surface trash showing between the furrow slices. Setting the skimmer too deep can prevent the plough working at the required depth, especially in hard land.

The disc coulter should be positioned approximately 12 mm (½in) towards the unploughed side of the share.

Front Furrow Width

Front furrow width is adjusted by rotating the cross shaft a few degrees. Depending on the model of plough, this is done either with a hand lever or by slackening the retaining bolts and rotating the cross shaft slightly with a spanner.

The skimmer point should be set just deep enough to turn the top corner of the furrow slice into the previous furrow bottom.

Ploughing Depth

Ploughing depth is controlled either with a depth wheel running on unploughed land or by the hydraulic draft control system on the tractor. Draft control is usually sufficient to maintain the required depth for commercial ploughing but a depth wheel will be more effective on multi-furrow ploughs and more accurate on a competition plough.

Levelling

The adjusting handle on the tractor's right-hand lift rod is used to level the plough and set all of the furrows at the same depth.

Pitch

Pitch refers to the angle of penetration of the shares. The pitch of the complete plough is adjusted by altering the length of the tractor top link. When correctly adjusted the plough will be seen to run level with the rear landside, making no more than a light mark on the furrow bottom.

A plough has too much pitch when the front body digs in deeper than the rear body. The ploughing will be uneven, the rear landside and heel iron will ride clear of the furrow bottom and there will be undue wear on the shares, especially when working in hard ground. Lengthening the top link will correct this fault.

A plough will have insufficient pitch when the front share tends to ride out and the rear landside makes a deep mark in the furrow bottom. This makes it difficult to plough at the required depth but shortening the top link will solve the problem.

Each body should have the same pitch measurement. This can be checked by measuring the vertical distance from an identical position on the underside of each beam to the tip of a new share. Some ploughs have an individual pitch adjustment on each body.

TRAILED CONVENTIONAL PLOUGHS

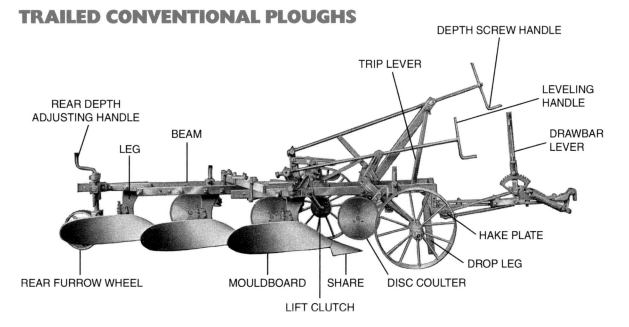

The main parts of a conventional trailed plough.

A trailed plough with three wheels and an adjustable drawbar has the same soil-engaging parts set up in the same way as on a mounted plough. Other field adjustments are made with the plough drawbar and levers or screw handles connected to cranked axles on the land and furrow wheels. A rope-operated mechanical lift clutch or rack lift mechanism on the land wheel of the plough lifts and lowers the bodies into and out of work.

The rack lift mechanism on the land wheel of a trailed plough.

Controls and Adjustments

Ploughing Depth

Ploughing depth is adjusted by the depth control screw handle or hand lever which raises and lowers both sides of the plough. On some ploughs the depth control handle also raises or lowers the rear furrow wheel to keep the plough level from front to back. Other ploughs have a separate depth-adjusting screw handle for the rear wheel.

The depth-adjusting handle on the rear castor wheel.

Levelling

Levelling is done with a screw handle or lever connected to the furrow wheel of the plough. It is used to tilt the plough when the front body ploughs a deeper or shallower front furrow than the one turned by the rear body. Uneven furrow depth will show up in the work as alternating high and low furrow slices.

A trailed plough drawbar. The angled bar is used to change the position of the plough in relation to the tractor wheels, while the hand lever provides a fine adjustment in each pinhole position.

The Drawbar

Vertical and horizontal settings on the plough drawbar are used to alter the width of the front furrow, adjust the pitch of the shares and hitch the plough so that its line of draft is central to the tractor.

Plough Pitch

Plough pitch relates to the angle of penetration of the shares. When correctly set the plough bodies will turn furrows of equal depth and size. With too little pitch it is difficult to get the front body to plough at the required depth and with too much pitch the front body will dig in too deep.

The sets of holes in the hake plates are used to adjust the pitch of the plough bodies.

Measuring the vertical distance from the underside of the beam to the tip of a new share is a simple way to check that each body has the same amount of pitch. Some ploughs have individual pitch adjustment for each body.

The pitch of a trailed plough is adjusted by altering the position of the hake bar on the plough frame. In normal ploughing conditions the pitch will be correct when the plough drawbar slopes slightly up towards the tractor. When the hake bar is too low the plough has insufficient pitch, the tractor tends to pull the front of the plough up out of work and the rear landside, or rear furrow wheel, leaves a deep mark in the furrow bottom. When the hake bar is too high there will be too much pitch, causing the front body to plough too deep while the rear landside runs clear of the furrow bottom.

Some trailed ploughs have a screw handle on the rear wheel. This will also be used to set the plough level from front to back when adjusting the pitch of the plough. Others have the rear wheel linked to the plough depth control lever with no separate adjustment for the castor wheel.

Front Furrow Width

Front furrow width is controlled by the horizontal position of the hake bar on the plough frame. It is important to hitch the plough to the tractor so that the main drawbar is parallel to the furrow wall, putting the line of draft of the plough as near as possible to the centre line of the tractor. When hitching the plough it should be attached so that the front furrow is at the same width as the other furrows.

If the rear of the plough swings or 'crabs' to the left or to the right the hake bar should be moved in the opposite direction to the 'crabbing motion'. Moving the hake bar to the left or to the right must be done without changing the

relative position of the tractor and the plough. Some ploughs have a hitch adjustment lever on the drawbar used to make minor alterations to the line of draft.

REVERSIBLE PLOUGHS

The tractor must be prepared before attaching a reversible plough to the hydraulic linkage. Enough front weight should be added so that with the plough in the raised position the tractor is stable. The rear wheels should be set at the track centre recommended in the plough book, usually with the inside of the tyre walls on the front wheels in line with the inside face of the rear tyre walls. The tyre pressure on each axle should be the same. Relatively small differences in tyre pressure, especially in the rear tyres, can affect ploughing depth by an inch or more. The hydraulic linkage lift rods should both be set at the same length as unequal length lift rods can affect the furrow depth on alternate runs. The mouldboards, shares, landsides, disc coulters and skim coulters on a reversible plough serve the same purpose and are adjusted in the same way as described earlier (see page 9) for conventional mounted ploughs. However, as the reversible plough has left- and right-handed bodies, coulters and skimmers it is critical to match the settings on both sides of the plough so that both sets of bodies turn equal-sized furrows.

Hitching

It is important to check that the cross shaft is central in the plough frame so that the top link and the plough run in line behind the tractor. Attaching the left lift arm followed by the right lift arm and then the top link is the best way to hitch the plough to a tractor which is not equipped with quick-attach linkage. The initial top link adjustment - setting it a couple of turns longer than the slack position - can be done with the plough standing on a hard surface.

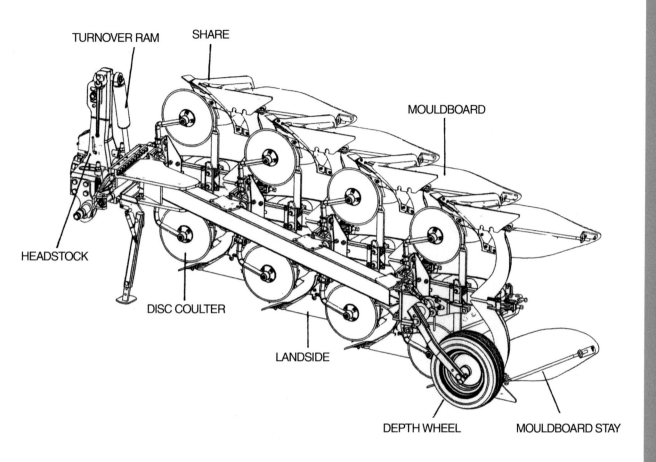

TURNOVER RAM   SHARE

MOULDBOARD

HEADSTOCK

DISC COULTER

LANDSIDE

DEPTH WHEEL   MOULDBOARD STAY

The main components of a reversible plough.

Front Furrow Width

Depending on the type of plough, front furrow width may be adjusted with a hydraulic ram between the head stock and plough frame, an adjuster screw on the turnover ram or front furrow width adjusting bolts on the headstock. The left lift rod levelling box can be used to fine tune front furrow width. The front furrow will be the correct width when it matches the furrow turned by the rear body on the previous run.

Depth

Depth is almost always controlled with the plough depth wheel or wheels but when ploughing in very wet or tough land the use of

the hydraulic draft control system can improve traction and reduce wheel slip.

The Pitch

The pitch of the plough is adjusted with the top link. If it is too short the front furrow will be too deep; if the top link is too long the rear furrow will be much deeper than the front furrow and the plough will have difficulty entering the ground.

Levelling

The plough is initially set with adjusters on the headstock and fine tuned with small changes to the length of the right-hand lift arm.

Articulated reversible ploughs have wheels part way along the frame to reduce the turning circle at the headland.

PLOUGH ALIGNMENT

Good ploughing cannot be done if the plough has damaged or bent components, possibly caused by the striking of a tree root or other hidden obstruction. Misalignment of one or more bodies may not be visually obvious and readjusting the plough may not overcome the problem. The following series of checks and measurements can be used to locate both vertical and lateral misalignment of a plough. They should be made after fitting a new set of shares and, where appropriate, wing shares. It is also important at this stage to check that the nuts and bolts securing the soil-engaging parts to the frame of plough are tight. Before taking measurements the plough should be parked on a flat and level surface, using the adjustments provided to set the plough frame level both lengthways and crossways.

Vertical Alignment

The pitch measurement should be the same for all bodies. This is the vertical distance from the underside of the beam to the tip of a new share. Measure with the tape held vertically from exactly the same position on each beam to the tip of the share. When the pitch measurement is not the same for each body this may be due to a bent plough beam. Some ploughs have a screw adjuster which can be used to reset individual body pitch at the correct

The pitch measurement must be the same for each plough body.

measurement. It is usual for reversible ploughs to have pitch adjustment on one set of bodies to facilitate matching the pitch measurement with the opposite set.

A plough with three or more bodies may have the correct pitch measurement but if the plough has a bent beam the tips of the shares will not be in line. Placing a straight edge alongside the share points is a simple way to check that the points are level and in line with each other. If they are not then the beam carrying the misaligned share will be bent.

Use a straight edge to check the shares are level and in line.

Using a straight edge to check the plough will turn equal width furrows.

With a two-furrow plough, a rough check of beam alignment can be made by placing two straight edges across the beams. If the tops of the straight edges are not in line it is likely that the frame is out of true alignment.

Lateral Alignment

A plough with a twisted or bent leg or beam will turn one or more unequal width furrows. The relative position of each body on its beam can be checked with a plumb line. This should be suspended from exactly the same position on each beam and it should touch the share in a position which is exactly the same distance from the landside edge on each share.

A straight edge positioned alongside the rear body and running forwards to the front of the plough can be used to check that the bodies, fitted with

new shares, are parallel with each other. The bodies will be parallel when the measurement from the straight edge to the tip of the share is the same for each plough body. An alternative method, suitable for a two-furrow plough, requires the removal of the shares. Straight edges projecting forward and placed alongside landsides should be parallel; if they are not it is likely that one of the beams or legs is twisted or bent.

Correcting Misalignment

Some ploughs have a pitch adjustment on each body. On ploughs without this luxury it is sometimes possible to correct small errors by slackening the beam or leg-retaining bolts and inserting a small wedge to raise or lower the offending body slightly.

A slightly bent beam can sometimes be pulled

back into alignment by clamping a strong brace across the top of the frame. Then, after slacking off appropriate bolts, it may be possible to pull the offending beam back into position before tightening the bolts again and removing the brace bar.

Washers or thin packing pieces can sometimes be placed between the leg and the beam to twist the body slightly sideways to set the bodies to plough equal width furrows.

Plough Bodies

A plough body must have sufficient 'suck' and 'lead to land'. Suck is necessary for the body to maintain its working depth and lead to land helps to maintain the correct furrow width. There will be too little suck when the plough has worn shares and it will be difficult to plough at the required depth, especially in hard ground. Worn shares and landsides will likewise reduce the lead to land and cause problems in maintaining the correct furrow width.

The mouldboards will not turn uniform-shaped furrows if they are not parallel with each other. Some ploughs have adjustable mouldboard stays, which are used to set the mouldboards parallel to each other. On ploughs with fixed mouldboard stays, inserting or removing one or more washers between the mouldboard and its stay will correct minor inaccuracies.

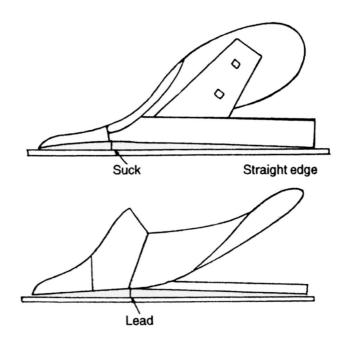

All plough bodies require adequate 'suck' and 'lead to land' - checked with a straight edge - in order to plough furrows at the required depth and width.

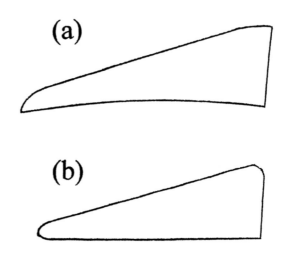

New (a) and worn (b) plough shares. A plough body with a worn share will have insufficient suck and lead to land, making it difficult to maintain the required furrow width and depth, especially when ploughing up hard land.

The first step in checking that the mouldboards are parallel is to make a measured mark in the same position on the top of the leading edge of each mouldboard. Then make a second measured mark on the top edge near the back of each mouldboard. The front and back marks must be the same distance apart on each mouldboard.

Measuring the distance between the marks at the leading edge of adjacent mouldboards. These should be equal on ploughs with three or more bodies.

The distance between the marks at the back of adjacent mouldboards should be the same as that between the marks at the leading edge of the mouldboards.

I. Types of Plough

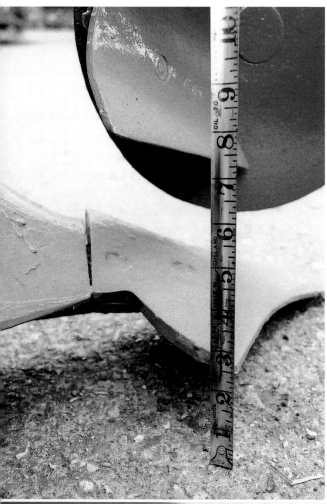

With the plough standing level the distance between the wing edge and the ground should be equal for all of the bodies. A bent or twisted leg or plough frame may be the reason for an unequal measurement between the wing and the ground.

The mouldboard extensions, or tailpieces, must be set at the same height on all bodies. They can be adjusted using the slot at the front which sometimes has a choice of two or three hole positions for the second retaining bolt.

2. Basic Ploughing Techniques

It takes time and practice to become good enough to win prizes even at local ploughing matches, and a great deal more practice will be needed before a novice can hope to qualify for national ploughing matches.

Learning to plough almost invariably starts on the farm, ploughing in stubbles or after clearing roots or other arable crops. Commercial farmland is almost invariably ploughed with a mounted or semi-mounted reversible model to the virtual exclusion of the conventional right-handed plough. Nonetheless, success at ploughing matches with a conventional plough will be easier to achieve with plenty of practical commercial experience. The following paragraphs, meant for the novice, aim to explain the basic techniques of conventional and reversible ploughing.

CONVENTIONAL PLOUGHING

Ploughing with a right-handed plough is usually done in sections, or lands. This is more usually known as systematic ploughing or ridge and furrow work. However, this system leaves ridges and troughs across the width of the field which is not desirable for the sowing and after cultivations of sugar beet and other root crops. An alternative system of round and round ploughing, working either from the centre to the headlands or vice versa leaves a relatively level field surface. Round and round ploughing is more suited to regular shaped fields with the best results obtained when ploughing from the centre of the field to the headlands.

Systematic ploughing

When ploughing in lands the first task is to mark out headlands and side lands with a shallow scratch furrow, which if turned towards the centre of the field will give neater ends to the ploughing. If the land is hard, turning the headland furrows towards the hedge will make it easier to get the plough to enter into the ground at the start of the run. Headland width depends on the size of plough, 9 m (9 yd) is about right for a three-furrow plough with an extra 1 m (1 yd) added for each additional furrow.

The next task is to divide the field into sections, or lands, and at the same time plough the first run of an arable opening split at intervals across the field. This is normally done across the shortest field dimension in order to create longer ploughing runs and a reduced number of headland turns. Carefully measured marking poles should be set out so that each opening furrow is parallel with its neighbour. The distance between each opening furrow will depend on the size of the plough. A reasonable spacing for the opening furrows might be 33 m (33 yd) for a two-furrow plough and 44 m (44 yd) for a three-furrow plough.

An arable split opening is not suitable for ploughing grassland, because the following cultivations are likely to pull some of the turf back to the surface. An alternative grassland opening is made by turning a shallow furrow with the rear body. On the second run the front body turns a shallow furrow, partly covering the first marking furrow with the remaining bodies turning slightly deeper. The ridge is completed on the third run with the plough set a little deeper and the front furrow covering most of the shallow furrow slice turned on the first run.

2. Basic Ploughing Techniques

With the first opening furrow of each ridge ploughed a four-furrow start or ridge is ploughed around each opening split. Ploughing can then get under way by alternately casting (ploughing between two furrows) and gathering (ploughing around a ridge) until the unploughed ground of each land is reduced to the correct width for the finish.

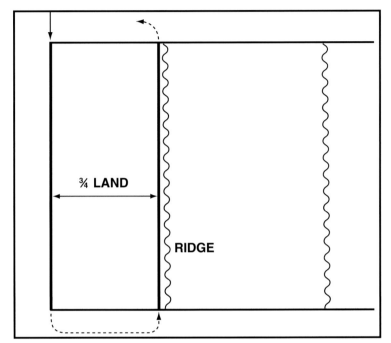

With the plough set for the first run of an opening split (see page 35) the first furrow is ploughed at a distance equal to three quarters of a full land measured from the headland marking furrow. When, for example, a three-furrow plough is used, the first land should be 31 m (33 yd) wide with the remaining lands 44 m (44 yd) apart. To avoid wasting time resetting the plough several times the remaining opening split furrows, each one carefully measured from its neighbour, can be ploughed at this stage.

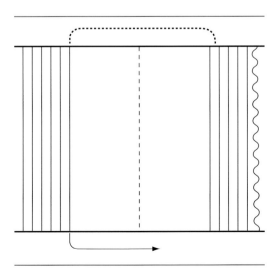

Casting – ploughing between two ridges.

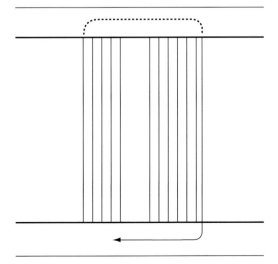

Gathering – ploughing round and round a ridge.

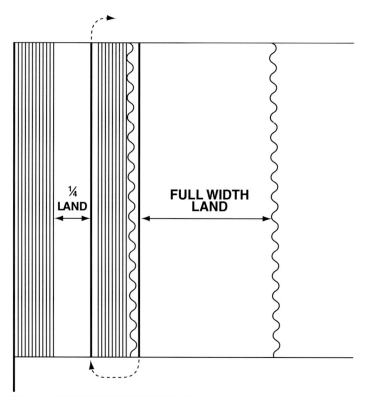

The first stage in systematic ploughing is to plough the three-quarter width land by casting between the headland mark and the first ridge until one quarter of this land is left to plough.

¼
LAND

**FULL WIDTH
LAND**

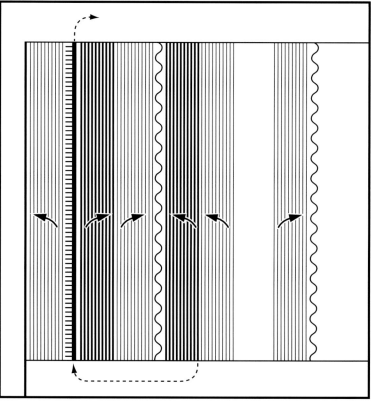

The remaining part of the three-quarter width land is ploughed by working round and round the ridge until, depending on the size of the plough, it has been reduced to the five or six furrow widths required to plough the finish. Ploughing between the first and second ridges is next, and this continues until one-quarter of the first full width land is left unploughed.

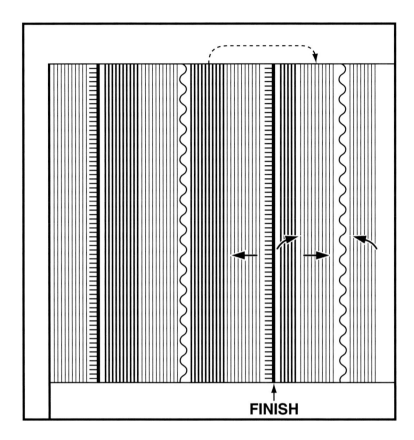

FINISH

The remaining quarter is ploughed by gathering around the second ridge until it is just wide enough for the second finish. This procedure is repeated until all the land between each ridge has been reduced to the last few furrow widths.

Each finish can be ploughed as work progresses across the field. However, less time will be spent in adjusting the plough to make the finish if this task is left until the land between each ridge has been reduced to the last five or six furrow widths.

Alternating the direction of ploughing the headlands – towards and then away from the hedge in alternate years – will keep the headland strip level and avoid creating a deep furrow around the edge of the field.
When the field is next ploughed it will help to keep it reasonably level by making the openings in the low ground left by the previous year's finishes with the new finishes coinciding with the previous year's opening ridges.

REVERSIBLE PLOUGHING

The disc coulters, skimmers, mouldboards and other soil-wearing parts of a reversible plough are set up in the same way as they are on a conventional right-handed plough but it is important to make sure that the same adjustments are made to both sets of bodies. Marking out a field for reversible ploughing is limited to making a headland mark at each end of the field which should be ploughed in the direction which reduces short work to a minimum. Ploughing in the longest direction will reduce the number of headland turns.
Headlands need to be wide enough to turn the tractor with the minimum of effort and the

width related to the tractor's turning circle and size of plough. Ten metres (30 ft) is about right for a three-furrow plough but much wider headlands are needed for turning with multi-furrow ploughs. Some tractor drivers make scratch marking furrows of the same width as the headland marking furrows on both sides of the field and after ploughing the main part of the field the side lands and headlands are ploughed round and round.

Start at one side of the field, either at the hedge or at the side land. Scratch furrow first, by ploughing one run away from the hedge to move the soil. Then plough in the opposite direction to turn the cut soil back towards the hedge and continue ploughing across the field. Plough out any short work as necessary before ploughing the headlands. Plough from an opposite side of the field each year and remember that ploughing the headlands towards and then away from the hedge will help to keep the field level.

BASIC PLOUGH SETTING FAULTS

Poor penetration
Provided the land is not too hard or dry difficulty may be experienced in getting the plough to work at the required depth because the shares are worn or the disc and/or skim coulters are set too low. A mounted plough will not plough deep enough if the top link is too long and the same problem will occur if the drawbar of a trailed plough is set too low on the hake plate.

The plough runs too deep
This can be a problem on soft or very wet land when the plough depth wheel follows the tractor rear wheel. A mounted plough will run too deep if the top link is too short; it should be adjusted so that the heel of the rear landside makes a light mark in the furrow bottom.

Uneven width furrows
When the front furrow is too narrow or too wide when using a mounted plough, the wheel track may not be at the recommended setting for the furrow width or the cross shaft may be incorrectly adjusted. When other furrows are of unequal width the disc coulters or skim coulters may require adjustment. There may also be incorrect spacing between the beams or legs.

Uneven furrows
Unequal furrow widths can cause uneven furrows. Other possible causes include too little or too much pitch, the plough does not run level or the mouldboards are not parallel with each other. Uneven wear of the shares, unequal disc coulter and skim coulter settings and unequal individual body pitch can also result in uneven furrows. A narrow rear furrow may be due to a wide rear tractor tyre disturbing the furrow slice. If a particular adjustment does not solve the problem then return that item to its original setting before trying another adjustment.

Furrows standing on edge
This can be due to the furrow being too deep for its width or the mouldboards being set too wide. When ploughing in grassland these faults can even result in the furrow slice rolling back into the furrow bottom.

This is a good example of the discs being set too low in very hard conditions, which resulted in an inability to control the movement of the plough.

This reversible finish is very untidy. The wrong width of land has been left for the final furrows, which has resulted in poor furrows and also leaves unploughed land next to the first furrow of the plot. This shows how important it is to measure accurately to leave the correct number of furrows in the finish.

The furrows in this picture are in pairs with the back furrow on top of the front furrow. This is a prime example of the top link being set too long.

3. Setting Out the Plot and Preparing the Tractor and Plough

Setting out the plot

The organisers of a ploughing match are responsible for the preparation of the site, accurately measuring out the plots and ploughing scratch furrows towards the headland at both ends of the plot. A marking peg, which provides the starting point for setting out the sighting poles, is placed in the headland marking furrow at each end of the plot.

Conventional ploughing plots are arranged so that the opening furrow is drawn at right angles to the headland scratch furrows. The plots in the reversible class are marked out so that the opening furrow is ploughed at an angle to the headland marks. Competitors are allowed to spread the soil from the headland scratch furrow to help them make tidy ins and outs.

Three sighting poles are allowed when drawing the first furrow. A fourth pole may be used at the near headland while setting out the poles but it must be removed before positioning the tractor at the headland for the opening run. The sighting poles are set out in exactly the same way for the opening run in all classes of competition ploughing. Some competitors have poles of the same length while others use a longer pole at the far headland. The three poles should be dead in line and absolutely vertical. Assistance with setting the poles out and removing them as the tractor progresses across the field is permitted.

The sighting pole is in the peg mark made by the match organisers.
To help make tidy ins and outs, competitors are allowed to spread the soil from both headland marking furrows.

Competitors may use an assistant to help line up the sighting poles.

Preparing the tractor and plough

A few basic checks and adjustments made in the farmyard can save both time and temper when opening up a plot at a ploughing match.

The Tractor

Most competitors use a 30 cm (12 in) furrow plough in the vintage and classic classes and 32cm (13 in) in the World-style conventional class with the tractor rear wheels set at 56 in (142 cm) centres to give a distance of 43–44 in (109–112 cm) between the tyre walls. This setting allows the tractor wheels to span the last three furrow widths of undisturbed ground when ploughing the finish.

The front wheels should be set with the insides of the tyre walls in line with the

The rear wheels should be set at 56 in (142 cm) for 30 cm (12 in) work and at 52 in (132 cm) centres when using a 25 cm (10 in) furrow plough.

inside walls of the rear tyres. When using a 25 cm (10 in) furrow plough the rear tractor wheel centres need to be at 52 in (132 cm) centres, again with the inside walls of the front tyres in line with the inside walls of the rear tyres.

Competitors in both the World-type conventional and reversible ploughing classes are allowed to use a slotted or hydraulic quick-entry top link and a hydraulically operated left-hand lift rod. A hydraulic ram on the right-hand lift arm for moving the plough sideways is also permitted in the World-style conventional and reversible classes. A quick-entry top link is not allowed when ploughing with a vintage mounted plough.

Tractors in production before 31st December 1959 are eligible for the trailed and mounted vintage classes and those in production before 31st December 1976 qualify for the classic and classic reversible classes. Tractors used in all ploughing match classes must comply with the current safety regulations.

The Plough

Ploughs used in World-style conventional and reversible ploughing classes can be equipped with various quick adjustments that will not be found on equivalent commercial ploughs. These include screw handles to alter individual furrow depth and disc coulter

Competitors in the vintage and classic mounted ploughing classes may not use a hydraulic top link or a hydraulically adjusted left lift rod.

Slotted or quick-entry top links are allowed in all mounted ploughing classes.

height, multiple tailpiece adjustments and spacers which can be inserted between the plough beams to fine tune the width of the furrows to suit the soil conditions.

In the vintage and classic classes at a ploughing match the bodies used must have been available when the plough was manufactured. Although the basic plough must be of the original design, competitors are allowed to make modifications that enable the plough to be adjusted more easily.

Bolt holes in the leg and leg bracket of this World-style conventional plough are used to vary the depth of the front furrow. The screw handle is used to make minor alterations to the depth.

Examples of permitted modifications for vintage and classic mounted ploughs include a second depth control wheel at the front, an axle or frame extension to reposition the depth wheel to make it follow the tractor wheel when ploughing the finish and a screw adjuster to move the plough sideways on the cross shaft.

Permitted modifications to vintage trailed ploughs include a furrow wheel axle extension to make it follow the tractor wheel when completing the finish.

In order to turn equal depth furrows the plough bodies must all have the same pitch measurement. This is the vertical distance from the tip of a new point or share to an identical point on the underside of each

Competitors are allowed to fit an extension to the depth wheel bracket on a vintage mounted plough so that the wheel can be moved out to follow the tractor wheel when ploughing the last furrow of the finish.

The mouldboards should be parallel throughout their length. To check this make two measured marks in identical positions on the top edge of each mouldboard and then measure the distance between these marks on adjacent mouldboards. If the measurements are not equal use the adjustable mouldboard stays to correct the error.

beam. Some plough bodies have an individual pitch adjustment, others may need more drastic treatment in the workshop.

For normal ploughing the disc coulters should be set with the bottom of the disc about 12 mm (½in) above the share and the same distance towards the unploughed land. The disc will need to be lowered below the share when making the initial opening run and for the finishing furrow. The skimmer should be set to run at no more than one third of the ploughing depth. Ploughing depth should be adjusted with the plough depth wheel; the tractor's hydraulic depth control system should only be used in extremely wet conditions.

A second rear depth wheel, permitted on World-type conventional ploughs, can be a useful aid on unlevel ground as it will maintain the correct ploughing depth if the main depth wheel drops into a furrow.

Ploughing Match Classes held under the rules of The Society of Ploughmen

I. Vintage
For tractors in production before 31st December 1959 using a plough equipped with bodies that were in production when the tractor was made.

The vintage division is divided into two classes of ploughs: (a) trailed, (b) mounted.

2. Classic and classic reversible
For tractors and mounted ploughs both in production before 31st December 1976.

3. High cut or oat seed furrow ploughing
Mostly done with trailed ploughs. This is the only class in which competitors may use boats, seamers and press wheels to improve the shape and appearance of turned furrow slices.

4. World-style conventional
For tractors using modern right-handed ploughs. N.B. Rules for world-style ploughing vary slightly in different countries.

5. World-style reversible
Many competitors use specially designed competition ploughs which allow the setting to be fine tuned. Competitors may also use the tractor hydraulic system's auxiliary services to fine tune the plough settings.

4. Vintage and Classic Mounted Ploughs

A typical mounted plough suitable for both vintage and classic competition classes. The class depends on the age of the tractor: those in production before 31st December 1959 qualify for vintage and tractors in production before 31st December 1976 qualify as classic.

The Double Split Opening

The double split opening forms the first two runs of a 12-furrow start with a two-furrow plough. In wet conditions it can be an advantage to plough an 11-furrow start (page 40) with the first two runs ploughed in the same way as they are for a 12- furrow start.

Twelve Furrow Start

The First Run

This is ploughed with the rear body only, turning a shallow furrow 6-8 cm (2-3 in) deep. The furrow will need to be a little deeper where there are tramlines or deep wheel marks to help keep the plough in line with the tractor.

The front body, or bodies, need to be lifted clear of the ground with the right lift rod levelling box and, where necessary, by lengthening the top link. Lower the depth wheel to set the rear body at the required depth for the opening furrow and lower the disc coulter to run alongside and close to the share to give a clean cut to the furrow wall.

The front body must be lifted clear of the ground with the right-hand lift rod levelling box before turning the first opening furrow.

It is a good idea to keep a record of the position of the depth screw handle and other settings used on this and following runs of the opening. It is important to make sure that the share cuts all the ground on the first opening run.

Careful driving is vital. Line up the centre of the tractor bonnet with the sighting poles so that only one of them can be seen. Keep the tractor dead in line with the poles and always stop before looking back at the work. When leaving the seat to remove a sighting pole it is important to take up exactly the same position on the seat before moving off.

The vintage mounted plough prepared for the first opening run with the rear disc lowered below the share and the front body raised well clear of the ground.

Ploughing a shallow opening furrow with the rear body.

An assistant is allowed to remove the sighting poles while the opening run is being ploughed.

The Second Run

The aim is to plough two dead straight furrows about 8 cm (3 in) deep. Before setting off on the second run the front body needs to be lowered using the handle on the lift rod levelling box. The depth wheel should be raised slightly to put the share in a little deeper to hold the plough in position with the rear disc running alongside the furrow wall from the first run.

Stop after ploughing a short distance to check the work and make any adjustment to the plough. Remember to stop before looking back at the ploughing while on this run. Checking the work while on the move is almost guaranteed to put a kink in the furrow.

On the second run the front disc coulter must run alongside the opening furrow wall, leaving no land unploughed in the centre of the opening split.

On completion of the second run of the split the judges will be looking for straight, well-cut furrows with all of the soil moved and furrow slices which are uniform throughout their length.

The wheel mark left by the plough wheel on the first run can be used as a driving guide for the front right tractor wheel.

The Third Run

After the opening has been judged some competitors turn a small single furrow from the bottom of the furrow ploughed on the first run into the middle of the opening before ploughing the first four full furrows of the start. This additional run will help to produce a start with four uniform and level furrows.

Ploughing a single small furrow into the opening using the rear body with the disc coulter running alongside the furrow wall from the first run.

The Fourth and Fifth Runs

The plough is put back to its normal settings but not to the full 15 cm (6 in) depth required for the main body of vintage hydraulic tractor ploughing plots or to the full 17cm (6½ in) ploughing depth for classic vintage ploughing plots.

Reset the coulter with the bottom of the disc just above and about 12mm (½ in) outside the share. It will need to be a little higher in hard ground. Shortening the rear mouldboard stay will leave a little more space for the tractor wheel on the fourth run and avoid a tell-tale tyre mark that may lose a few points.

The front right wheel needs to follow the mark left by the tractor rear wheel on the previous run while ploughing the first two furrows of the start.

No further adjustments should be necessary before ploughing the next two furrows to complete the four-furrow crown which forms the first part of a 12-furrow start.
On this run two full-width furrows at about three quarters of the 15 cm (6 in) minimum depth required in the vintage class are ploughed back into the centre of the opening split.

The front right wheel runs in the normal ploughing position against the furrow wall while ploughing the third and fourth furrows of the start.

Completing the first four furrows or crown of the 12-furrow start.

Completing the 12-Furrow Start

Ploughing round the first four furrows, or gathering, continues until 12 furrows have been ploughed with the plough set to at least the minimum depth of 15 cm (6 in) required in the vintage ploughing class. Provided that the first four furrows were ploughed at shallower depth the start should have 12 level and equal-sized furrows.

The start is now ready for judging. The highest number of points will be awarded for a start with furrows that are uniform and straight throughout their length. The centre furrows should be closed for good weed control and there should be no high, low or narrow furrows.

The 11-Furrow Start

An 11-furrow start can be an advantage in wet conditions. Ploughing a very shallow front furrow and a full-depth rear furrow on the third run will make it easier to pull the plough. On sloping ground it will also be an advantage to plough uphill on this run.

Apart from a few high points on the first four furrows caused by some very large stones this 12-furrow start is level across its width.

The sighting pole shows that points will be lost for the high furrows at the centre of this start and still more will be lost for the open centre with unburied trash.

The first two runs of an 11-furrow opening split are ploughed in exactly the same way as explained for a 12-furrow start.

The right tractor wheel needs to run in the furrow bottom, approximately 25–30 cm (10-12 in) further over than it would be when ploughing the third run of a 12-furrow start. The rear body ploughs where the front body would run on a 12-furrow start.

On the third run of an 11-furrow start the front body turns a shallow furrow to support a full-width furrow ploughed by the rear body.

The Third Run (11-furrow start)

The depth wheel is raised and the plough levelled with the lift rod levelling box so that the front body turns a two-thirds depth furrow to support a full-depth furrow turned by the rear body.

The Fourth Run (11-furrow start)

For the next run the plough should be set to turn two full-depth furrows. The front tractor wheel needs to run in the furrow to press down the small furrow slice turned by the front body on the previous run. On reaching the headland the first three furrows of an 11-furrow start are complete.

The front right tractor wheel runs against the furrow wall on the fourth run of an 11-furrow start.

Turning two full-width furrows completes the first three furrows of the alternative 11-furrow start.

When the last two furrows of the 11-furrow start have been ploughed the judges will look for straight, even and uniform furrows that are level across the full width of the start.

Ploughing the Plot

When the 11- or 12-furrow start has been judged, the land is ploughed by casting between one side of your start and the start on the neighbouring plot with the next highest number.

Before starting to plough the main part of the plot it is important to measure the distance between the two furrow walls in at least three positions along the length of the plot to make sure the two furrows are parallel. The judges will ignore any narrow or wide furrows ploughed on the first two runs on the neighbour's side of the plot to correct any fault.

As well as checking that the furrow walls on both sides of the plot are parallel it is equally important to make sure that the correct width of unploughed land is left between the two furrows to achieve an accurate finish to the plot. There needs to be an odd number of furrow widths when using a 30 cm (12 in) plough but either an odd or even number of furrow widths can be left when working with a 25 cm (10 in) plough. Achieving the correct distance between the furrows is important to ensure that the tractor wheels will straddle the last three or four furrows left to plough when making the finish. The wheels are best set at 56 in centres (142 cm) for 32 cm (13 in) ploughing and 52 in centres (132 cm) for 25 cm (10 in) work.

Points will be lost if any ploughing is not at least at the minimum required depth, so it is important to check from time to time that the furrows are at least 15 cm (6 in) deep in the vintage mounted class and 17 cm (6½ in) deep in the class for classic mounted ploughs.

Ample time is allowed to plough a competition plot. Great care must be taken to keep the ends of the furrows as square as possible with the headland mark at each end of the plot. Carefully line up the tractor at the headland before lowering the plough into work.

The ends of these furrows are far from square with the headland mark so there will be little chance of scoring many points for straightness and a parallel finish.

As work progresses make regular measurements to check that the plough is working at the minimum required depth and that the two furrows remain parallel and are the correct distance apart. If the two furrows are not parallel or the wrong distance apart at this stage and it is necessary to turn a wide or narrow furrow to rectify the fault, then points will be lost.

When ploughing towards the finish it is very important to have exactly the right width of unploughed land so that the tractor wheels span the three or four furrow widths left to plough. If it is too wide or too narrow the land wheel of the tractor will either run on the very edge of the remaining unploughed ground and break down the furrow wall or drop into the furrow bottom. In both situations this fault will be obvious to the judges as there will be too little or too much land left to plough on the final run of the finish.

Ins And Outs

Points are awarded for clean entries (ins) and exits (outs) at the headland. Care should be taken to drop the plough in and lift it out at exactly the same position every time. Wind the front furrow in a little with the levelling box on the lift rod both when going into and when coming out of work at each end of the plot. The use of a quick-entry slotted top link with a vintage or classic mounted plough can help to achieve neat ins and outs when lowering the plough into the ground and when lifting it out of work at the headland.

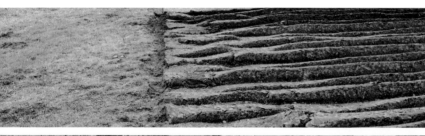

The judges would award a very high score for the ins on this plot. However, the handling, treading or shaping of furrows is not allowed except for oatseed furrow ploughing.

Very few points would be awarded for these ragged furrow ends.

The Finish

The finish consists of the last eight furrows with a 25 cm (10 in) or a 30 cm (12 in) plough. The judges will award most points for a good finish which will have eight uniform and straight furrows, which are gradually reduced in depth with a shallow and narrow final furrow. The tractor wheel mark should be level with the other furrows – points will be lost if it is too deep.

When ploughing the last eight furrows, each run should be ploughed at a lesser depth than the previous one to help achieve a level and shallow finish. Great care is needed at this stage to keep the ins and outs as square as possible with the headland marking furrows. Careful measuring will have made sure that when reducing the unploughed land from six to four furrow widths there will be enough space for the front left tractor wheel to run on unploughed ground.

Four Furrows Left To Plough

A number of adjustments must be made to the plough before ploughing the next two furrows. Repositioning the depth wheel behind the tractor's left wheel will help to keep the plough at an even depth. On ploughs with a depth wheel that cannot be moved sideways it may be necessary to control furrow depth with the hydraulic lift lever on this run.

Lengthening the right-hand lift rod and the top link will help to keep the plough level. Fine tuning of the lift rod and depth wheel adjustments will be needed at the start of the run so that both bodies turn shallower furrows than on the previous run with the front body the deeper of the two. Both wheels need to run in the bottom of the furrows with the tractor's left wheel close up against the furrow wall on this run.

With six furrows left to plough the depth wheel should be lowered a little so that both bodies plough slightly shallower furrows than on the previous run.

The unploughed land has been reduced to exactly four furrow widths so that both tractor front wheels are in their correct position in the furrow bottoms with the front left wheel running against the furrow wall.

Two full-width furrows are ploughed on this run leaving two more to plough to complete the finish.

The unploughed land on this plot is too wide, causing the front right tractor wheel to ride up out of the furrow. The judges are bound to see the wide front furrow being ploughed on this run.

Two Furrows Left To Plough

On arrival at the headland it is necessary to return to the opposite end of the plot to comply with the rule that requires the final furrow of the finish to be turned towards the competitor's own start.

A single furrow is ploughed on the next run so that one furrow width will be left to plough on the final run of the finish. To achieve this the plough must be moved across to the right of the cross shaft. In this position the front body will run in the furrow bottom while the rear body turns a single furrow, leaving one unploughed furrow width for the final run of the finish.

Ploughing match rules allow the use of a screw adjuster to move the plough across on the cross shaft. Without this luxury the cross shaft will have to be moved towards the left of the plough after slackening the retaining bolts.

Before ploughing a single shallow furrow on the last but one run the disc coulter on the rear body must be lowered so that it runs alongside the share. The depth wheel, still positioned behind the tractor's left wheel, needs to be lowered to set the rear body to plough a one-third to half-depth furrow. It is important to ensure the rear body is no deeper that this as there must be enough soil left in the furrow bottom for the rear body to turn a sole or earth furrow on the final run.

The front left tractor wheel must run alongside the furrow wall on this run. Great care is needed at this stage to plough the straightest possible furrow so always stop before looking back to check the work.

Ploughing a one-third to half-depth furrow with the rear body on the last but one run. The plough has been moved over to the right of the cross shaft with the plough depth wheel following the tractor's left wheel.

The front body runs in the previous furrow bottom while the rear body turns a one-third to half-depth furrow on the last but one run of the finish.

The Final Run

The full unploughed furrow width left on the previous run together with a sole furrow taken from the bottom of the previous furrow are ploughed on the final run of the finish.

The depth wheel still follows the tractor wheel so that it runs in the single wheel mark left on the neighbour's side of the finish. The front body needs to be lifted with the lift rod levelling box so that it turns a half-depth front furrow while the rear body turns a sole furrow about 8 cm (3 in) deep, taken from the bottom of the previously turned furrow.

The plough set for the final run with the plough wheel following the tractor's left wheel and the rear disc coulter dropped down below the share.

The tractor's front right wheel needs to run against the furrow wall on the final run. Only one wheel mark is allowed on the finished plot so the tractor's left wheel must follow the wheel mark made by the tractor's right wheel on the previous run.

The judges will look for a finish that is uniform throughout its length, shallow and straight. All of the land must be cut with only one wheel mark showing and the last eight furrows sloping gently down to the final, shallow open furrow.

Very few points will be awarded for this finish which is not very straight and in which some of the land has not been cut.

5. Trailed Vintage Ploughs

The plough is correctly hitched in a central position on the drawbar for the first run of the opening split.

The Double Split Opening

The double split opening forms the first two runs of a 12-furrow start with a two-furrow plough. In wet conditions an 11-furrow start (page 57) with the first two runs made in the same way as for a 12-furrow start can be an advantage.

Basic controls and settings for trailed ploughs are described on page 13. Although the plough can be set up on the headland, soil conditions will be

The front body has been raised clear of the ground, the rear disc coulter has been lowered and the rear body has been set to turn a shallow furrow on the first run of the opening split.

different in almost every field and minor adjustments are bound to be necessary after ploughing a short distance on the first run. The tractor wheels need to be set at 52 or 56 in (132 or 136 cm) centres with the plough hitched

centrally on the tractor drawbar. The method of setting out the sighting poles with or without an assistant for vintage trailed ploughing classes is explained on page 30.

Stop the tractor after driving a short distance with the plough turning a shallow furrow 6–8 cm (2-3 in) deep with the rear body. Make any necessary minor adjustments then, after taking up the same position on the tractor seat, plough a little further before stopping again to check all is well. It may be necessary to keep repeating this procedure until the rear body is turning an opening furrow of the correct shape and depth.

Twelve-Furrow Start

The First Run

Although the adjustments are made in a different way, a trailed plough is set in much the same way as a vintage mounted plough for the first run of the split. The front body is taken out of work with the levelling handle, the rear transport wheel is raised to put the weight of the plough on the rear body and the rear body is adjusted with the depth control handle to turn a furrow about 6–8 cm (2-3 in) deep. Finally, the rear disc coulter is lowered to run alongside the share.

With the plough prepared for the first run, position the tractor at the headland mark with the centre of the bonnet in line with the sighting poles so that only one of them can be seen when sitting comfortably on the tractor seat. Lower the plough into work and drive forward a short distance, then, as always, stop before looking back to check the work. It's likely that some fine tuning with the depth and levelling handles will be required at this stage.

The Second Run
(12-furrow start)

For the second run of the opening split, the plough must be levelled with the front body set to turn a furrow at the same depth as the furrow ploughed on the first run.

The front right tractor wheel should run in the mark left by the plough wheel on the first run with the front disc coulter and share running alongside the first furrow wall.

Careful driving, and always stopping before looking back at the work, should result in an opening split with all of the soil cut and furrows that are straight and uniform throughout their length.

With the first single shallow furrow ploughed, the mark made by the plough wheel on the first run is used as a driving guide while ploughing the second run of the opening split.

For the second run the plough is positioned with the front disc coulter and share running alongside the first opening furrow wall.

The Third Run (12-furrow start)

While the judges award points for the opening split there is plenty of time to put the plough back to its normal settings for the next two runs.

The rear disc coulter needs to be returned to its normal ploughing position with the bottom of the disc just above and 12 mm (½ in) outside the share. It may need to be set a little higher in hard conditions. Lengthening the rear mouldboard stay may create a little more space for the rear tractor wheel when ploughing the third and fourth runs of the start.

Set both bodies to plough at the same depth with the depth and levelling handles but do not set them down to the minimum depth required for the main part of the plot yet. A minimum depth of 15 cm (6 in) is required in the trailed vintage tractor ploughing class and 17 cm (6½ in) in the classic vintage ploughing classes.

The front body turns a shallow furrow of the same depth as the single furrow ploughed on the first run. The rear body turns a small furrow to support the front furrow to be ploughed on the next run.

The front right tractor wheel should run in the furrow bottom left by the rear body on the second run of the opening split when ploughing the first two furrows of a 12-furrow start.

The Fourth Run
(12-furrow start)

Before ploughing the next two furrows the only necessary adjustment is to return the mouldboard to its original position if it was lengthened on the previous run. The front right tractor wheel should run alongside the furrow slice turned by the front body on the previous run. In some conditions it may be necessary to shorten the front mouldboard stay to prevent pushing the front furrow over too far.

Ploughing the first two two-thirds depth furrows of a 12-furrow start.

Ploughing the second pair of furrows to complete the first four furrows of the start, making sure that no rear wheel tyre marks will be visible on the furrow slice.

The front right tractor wheel should run close up against the furrow slice turned by the front body on the previous run while ploughing the second run of a 12-furrow start.

Completing the 12-Furrow Start

Ploughing continues by gathering (ploughing round the first four furrows), until the 12-furrow start has been ploughed at the required minimum depth of 15 cm (6 in) with a vintage tractor and plough or 17 cm 6½ in) in the classic ploughing class.

Completing a 12-furrow start. The judges will look for straight and uniform furrows that are level across the full width of the start.

The 11-Furrow Start

The procedure for ploughing an alternative 11-furrow start, described on page 40 for vintage mounted ploughs, may also be used in the trailed ploughing classes.

The first two runs of an 11-furrow start are ploughed in the same way as a 12-furrow start but, before making the next run, the plough must be moved across on the tractor drawbar to plough a single furrow with the rear body. The front body runs light in the previous furrow bottom. The plough is moved back to a central position on the drawbar before ploughing the next two furrows to complete the opening for an alternative 11-furrow start.

Ploughing the Plot

When the 11- or 12-furrow start has been judged, the land is ploughed by casting between one side of your own start and the start on the neighbouring plot with the next highest number.

Before starting to plough the main part of the plot it is important to check that the two furrow walls are parallel with each other. This is done by taking measurements in at least three positions along the full length of the plot. It is equally important to make sure that the correct width of land is left between the two furrows to give an odd number of furrows to plough with a 30 cm (12 in) plough and either an odd or even number when ploughing 25 cm (10 in) wide furrows.

Ample time is allowed to plough a competition plot and as work progresses it is important to check that the two furrows remain parallel and the correct distance apart.

If the two furrows are not parallel or at the incorrect distance apart, either or both faults should be rectified while ploughing the first two runs on the neighbour's side of the plot. The judges will ignore these runs and no points will be lost. However, if at a later stage it becomes necessary to turn a wide or narrow furrow to rectify a fault, then points will be lost if the judges notice the correction.

Generally the tractor wheels are set at 56 in centres (142 cm) for 30 cm (12 in) ploughing and 52 in (132 cm) for 25 cm (10 in) work. It is important to ensure that the wheels will span the last three or four furrows when ploughing the finish.

Measuring ploughing depth from time to time should ensure that points will not be lost when the stewards make their random checks. Furrow depth must be a minimum of 15 cm (6 in) in the vintage class and 17 cm (6½ in) for classic ploughs.

Ins and Outs

Points are awarded for clean entries (ins) and outs (exits) at the headland. Care should be taken to drop the plough in and lift it out at exactly the same position on each run. Using the levelling handle to put the front body in a little deeper for a short distance as the plough goes into work and raising the front body slightly when the plough is almost at the end of the run will help to keep the ends of the plot neat and tidy.

Great care must also be taken to keep the ins and out as square as possible with the headland marking furrows. This can only be achieved by carefully lining up the tractor at the headland before lowering the plough into work and keeping the tractor on a straight line until the rear body is up to the headland mark.

The judges would probably award most of the points available for the ins and outs on this plot but some will be lost for the uneven furrow slices near the end of the plot.

Very few points would be awarded for these ragged furrow ends.

The Finish

The finish consists of the last eight furrows. The judges will look for eight uniform and straight furrows, which gradually reduce in depth with a narrow and shallow final furrow. Only one tractor wheel mark is allowed and this should be level with the adjoining furrows – points will be lost if this wheel mark is too deep.

Two or three turns of the depth control handle to reduce the depth on the first two runs of the finish should be the only necessary adjustment to the plough. Ploughing at this slightly reduced depth will help to achieve a good shallow and level finish. Do take great care at this stage to keep the ins and outs as square as possible with the headland marking furrows.

To achieve a good finish both furrows must be parallel for the full length of the plot and square with the headland mark. With bends at both ends of these furrows there is little chance of many points being awarded for the finish on this plot.

Ploughing continues at the minimum competition depth until eight furrow widths are left to plough.

To ensure a good-quality finish check furrow depth at this stage.

Six Furrows Left to Plough

The plough should be set slightly shallower than it was on the previous run. Careful measuring while ploughing the plot will have made sure that the land has been reduced to exactly six furrow widths along its entire length, leaving enough space for the tractor's left wheels to run on unploughed ground. If the unploughed land is too wide, or too narrow, the result will be no different from that shown on page 47 for a mounted plough.

Four Furrows Left to Plough

One last check with the tape measure will make sure that there are exactly four furrow widths left to plough. The next three furrows should be ploughed towards the neighbour's plot before turning the fourth and final furrow in the opposite direction, leaving a single wheel mark on the neighbour's side of the finish.

Ploughing two three-quarter depth furrows to reduce the unploughed land from six to four furrow widths.

It will be necessary to return to the opposite headland before ploughing the next run of the finish to ensure that the tractor wheel mark is left on the side of the finish next to the plot with the next highest number. The plough needs to be set to turn two furrows at about half depth with the front left tractor wheel running against the furrow wall.

If the furrows at each side of the unploughed strip of land are not parallel or the wrong distance apart, the front right wheel will either run too far across on the ploughed land or tend to ride up on to the unbroken ground.

Two Furrows Left to Plough

Again the tractor and plough must return to the opposite headland and on the return run the unploughed land will be reduced to one furrow width. The plough must be moved across to the right of the tractor drawbar so that the front furrow runs in the previous furrow bottom while ploughing a half-depth furrow with the rear body.

The exact drawbar position will depend on the tractor used. The most important factor is to leave one full furrow width to plough on the final run.

The front left wheel needs to run against the furrow wall while reducing the unploughed land to two furrow widths.

Ploughing two half-depth furrows to reduce the unploughed strip to exactly two furrow widths.

The plough needs to be hitched to the right of the tractor drawbar to turn a single shallow furrow with the rear body.

The front left wheel should run against the furrow wall on this run while ploughing a shallow full width furrow with the rear body.

Ploughing a shallow furrow with the rear body to reduce the unploughed land to a single furrow for the final run of the finish.

The front disc should run against the furrow wall on this run while ploughing a shallow full-width furrow with the rear body to leave the required single furrow width to plough to complete the finish.

A single furrow width is left to plough. To comply with the rules the right wheel mark will be covered by the furrow slice turned on the final run. The rear body turns a shallow earth furrow to provide a level finish.

The Final Furrow

Adjustments must be made before completing the final run. On ploughs where the depth wheel can be repositioned behind the tractor wheel this should be done so that the plough wheels follow in the tractor wheel mark so that the plough will maintain an even depth.

Before setting off on the final run the depth handle on the plough must be used to lift the rear body so that it ploughs a shallow earth or sole furrow taken from the furrow bottom. A few turns of the levelling handle will also be required to raise the front body so that it ploughs an approximately two-thirds depth furrow.

Repositioning the plough land wheel so that it follows the left tractor wheels while ploughing the last run of the finish.

The front right-hand wheel needs to run alongside the furrow wall on the final run.

The front body ploughs the last full-width furrow while the rear body turns an earth or sole furrow from the previous furrow bottom.

Ploughing the final furrow.

The judges will award high marks for this finish which is uniform through its length with straight and shallow furrows. All the land has been cut and the last eight furrows slope gently down to the final shallow open furrow.

The furrows of this finish, which are uneven and too deep, will not add many points to the total score for this plot.

6. World-Style Conventional Mounted Ploughs

Competition plots for modern right-handed ploughs, generally known as World-style conventional ploughs, are marked out with a headland scratch furrow turned outwards at each end of the plot. Competitors are allowed to spread the soil from the headland-marking furrows. Two alternative methods of lining up the sighting poles for the first run of the opening split are explained on page 30.

With a World-style conventional plough the start can have either 11 or 12 furrows; an 11-furrow start can be an advantage in wet conditions. A double opening split is ploughed on the first two runs then, after the split has been judged, ploughing continues until the full 11 or 12 furrows have been ploughed.

The 12-Furrow Start:

Double Split Opening – First Run

This is ploughed with the rear body turning a shallow furrow 6–8 cm (2-3 in) deep. The first furrow may need to be a little deeper where there are tramlines or deep wheel marks to help keep the plough in line with the tractor.

Shortening the right-hand lift rod will bring the front body well clear of the ground and it can sometimes be an advantage to remove the front share. As a general rule the top link needs to be slightly longer for the opening than it does for normal ploughing. To achieve a clean-cut furrow wall lower the rear disc coulter so that it runs just below and close to the side of the share.

When using tailpieces, a neater and flatter opening furrow slice may be achieved if the rear body tailpiece is set low.

The plough set for the first run with the front body lifted clear of the ground and the rear disc adjusted to run just below and close to the side of the share.

A hydraulic top link will make it easier to achieve neat ins and outs at the headlands.

An assistant can be used to remove the sighting poles.

Careful driving is essential at this and every other stage when ploughing a competition plot. The golden rule is never to look behind without stopping first. After getting off the tractor to remove a sighting pole make sure you are sitting in exactly the same position on the seat before moving off again.

Before ploughing the first furrow it is important to set the tractor on the headland with the centre of the tractor bonnet perfectly in line with the sighting poles. Drive forward a few feet then stop before checking that the rear body is turning the correct size and shape of furrow. If it is not right then make the necessary adjustment, drive on a short distance then stop again to check the work.

Ploughing the first furrow 6–8 cm (2-3 in) deep with the rear body. The tailpiece is set low to flatten the furrow slice.

Neat ins and outs are important in all classes of competition ploughing. Using either a hydraulic or a slotted top link – both are permitted with World-type conventional ploughs – will make it a simple matter to lower the front of the plough at the start of the run and then lift it slightly when both bodies are in work. It can also be used to raise the front of the plough a little before lifting it fully out of work at the headland mark.

The Second Run (12-furrow start)

The same golden rule – never look behind without stopping first – is equally important when ploughing the second run. The front body should plough a furrow about 8–10 cm (3-4 in) deep with the rear body turning a shallower earth furrow. The plough may need to run a little deeper where there are tramlines or deep wheel marks. If it can be done, moving the depth wheel to a forward position so that it runs in the furrow

made on the first opening run will help to plough a furrow of a more even depth.

Before setting of on the second run of the twelve-furrow start, adjust the right-hand lift rod levelling box to bring the plough almost level and raise the depth wheel a turn or two to turn slightly deeper furrows on the second opening run.

Shortening the rear mouldboard stay will overcome the problem of the rear furrow slice being pushed too far to the right and setting the front tailpiece in a low position will flatten the front furrow slice.

The front right wheel of the tractor needs to be positioned 30–35 cm (12-14 in) away from the disc cut made on the first run. Great care must be taken to drive the tractor with the front wheel on or close to the plough wheel mark, making sure that no land is left unploughed.

The mark made by the front depth wheel when ploughing the first furrow provides a driving guide for the second run of the opening split.

The Third Run (12-furrow start)

While the opening split is being judged the plough can be prepared for the third run of a twelve-furrow start. The plough is put back to its normal settings but not to the full minimum 20 cm (8 in) competition depth required for this type of plough. Reset the disc coulters with the bottom of the disc 12 mm (½ in) outside and just above the share. They will need to be a little higher in hard land. Shortening the rear mouldboard stay will leave a little more space for the tractor wheel on the next run.

Finally, lowering the tailpieces will press down the furrows ploughed on this run and help to leave a start level across its full width.

The front share must run alongside the furrow wall made on the previous run to make sure that no land is left unploughed. At the end of the second run all of the land should be cut, leaving an opening split with straight, uniform and well-turned furrows.

The front right wheel needs to run in the bottom of the furrow turned on the first run of the opening split while ploughing the first two furrows back into the split.

When ploughing in stubble it may be best to turn a (6–8 cm (2-3 in) deep chip furrow into the bottom of the opening split before ploughing the third run as described previously. This is done with the front right tractor wheel running in the furrow and the plough set in the same way as it was for the first run.

It can also be an advantage to plough a slightly wider rear furrow on the third and fourth runs to achieve a better match of furrows across the start. This can be done by setting the disc coulter about 12 mm (½ in) wider than normal or, when possible, by widening the plough frame.

With the plough reset for the third run, both bodies return the soil ploughed outwards on the previous furrow slices together with fresh ground back into the centre of the opening split.

The three-quarter depth furrows ploughed on the third run form the first two furrows of a 12-furrow start

The Fourth Run (12-furrow start)

No further adjustments should be necessary before ploughing the next two furrows, again at about three-quarter depth, to complete the four-furrow crown.

The second two furrows of the 12-furrow start are ploughed on the fourth run with the front right tractor wheel running against the furrow wall.

The skimmers are doing their work with no trash showing between the two three-quarter depth furrows being ploughed to complete the four-furrows of the crown.

The completed first eight furrows of this 12 furrow start are straight with even furrow slices. The furrows are well packed with all the grass and trash well buried, apart from the centre, where it could be closed a little better in places.

Completing The 12-Furrow Start

The plough must now be returned to its normal settings before the next eight furrows are ploughed. Reset the disc coulters for normal ploughing with the bottom of the disc running just above and 12 mm (½in) outside the share. Return the rear skimmer to its normal setting, lengthen the rear mouldboard stay until the mouldboard is parallel with the one on the front body and put the tailpiece, if used, back to its normal position.

The depth wheel must be raised by a couple of turns on the handle, as the plough must work at the minimum competition depth of 20 cm (8 in) for World-style ploughs. A tape measure is an essential item in a ploughing match competitor's toolkit. Frequent checks of furrow depth and width are important at this stage. If the plough is not deep enough points will be deducted for every centimetre of error.

Ploughing the next run of a 12-furrow start. It is important at this stage to keep the ins and outs as neat as possible and square with the headland marking furrow.

A completed start ready for judging. The sighting pole shows it to be level across the full width of 12 furrows.

When the last two furrows of the start have been ploughed the judges will be looking for straight and uniform furrows which are level across their width with the last two runs ploughed to the required minimum depth. Penalty points will be deducted for each centimetre of error in furrow depth below the minimum 20 cm (8 in).

The 11-Furrow Start

An 11-furrow start with a three-furrow crown can be an advantage in wet conditions. The first two runs of the opening split are ploughed in exactly the same way. Before ploughing the third run the depth wheel must be raised and the plough levelled with the lift rod levelling box so that the front body turns a small furrow to support a two-thirds depth furrow made by the rear body.

The tractor is driven with the front right wheel in the furrow bottom but approximately 25–30 cm (8-10 in) further over than it would be for a 12-furrow start. In other words the tractor is driven so that the rear body is in the same position as the front body would be when making a 12-furrow start.

Ploughing two two-thirds depth furrows to complete the first three furrows of an 11-furrow start.

The fourth run, which completes the first three furrows of the crown, is ploughed with the bodies set to turn two full-width and two-thirds depth furrows. The tractor is driven with the front right wheel running in the same position in the furrow bottom as it was on the third run.

The eight remaining furrows are ploughed in exactly the same way as for the more usual 12-furrow start. The last two furrows on each side of the start must be ploughed at the minimum required depth of 20 cm (8 in).

At this stage the judges will be looking for eleven straight, firm and uniform furrows which are level across the full width of the start.

Ploughing The Plot

When the start has been judged the next stage will be to plough out the plot by casting between one side of your own plot and the start on the neighbouring plot with the next highest number.

The distance between the furrow wall on each side of the land left to plough must be measured before ploughing the first run on the neighbour's side of the plot to make sure the two furrows are parallel. Measurements need to be taken in at least three positions along the length of the plot. As well as checking that the two furrows are parallel you must check that the width of land left between them to be ploughed is equal to the exact width of the furrows left to plough. If the two furrows are not parallel or not at the correct distance apart then either or both of these faults should be corrected while ploughing the first two runs on the neighbour's side of the plot.

The judges will ignore the first four furrows against the neighbour's plot if at this stage they are ploughed to make the furrows at each side of the plot parallel. No points will be lost here but if it later becomes necessary to plough a wide or narrow furrow to overcome a problem, then points will be lost if the judges see the correction.

However, careful and regular measuring in at least at three places between the two furrows to make sure they remain parallel and at the correct distance apart should avoid the need to plough an irregular correcting furrow at a later stage.

Ins and Outs

Points will be awarded for clean ins and outs at the headland mark. Care should be taken to drop the plough in and lift it out in exactly the same position each time. It will improve matters if the front body is set in a little deeper, both when entering and when coming out of work at either end of the plot.

A hydraulic top link makes it easier to make quick entries and exits at the headland mark. The alternative mechanical quick-entry or slotted top link will also help to make regular and good-quality ins and outs.

The judges will award a high score for the good ins and outs on this plot.

Few of the points available for neat ins and outs would be awarded here.

It is equally important to keep the ins and outs as square as possible with the headland marking furrows. This is achieved by carefully lining up the tractor before lowering the plough into work.

Points will be lost if the ends of the furrows are not square with the headland mark.

The Finish

The finish consists of the last eight furrow widths of the plot. Ploughing must continue at the minimum required depth of 20 cm (8 in) until starting plough the finishing furrows. Making regular measurements while ploughing the plot should ensure that the furrows on each side of the plot remain parallel and the correct distance apart. This is important to ensure that there are exactly five or six furrow widths left to plough for the finish.

The judges will want to see eight uniform and straight furrows gradually reducing in depth with a shallow and narrow final furrow. The single tractor wheel mark should be level with the furrows – points will be lost if this wheel mark is too deep.

The only adjustment necessary before reducing the unploughed land down to five or six furrows will be to reduce the ploughing depth by 2–3 cm (½–1 in) to help achieve a level and shallow finish to the work. It may also be necessary to lower the skimmers. Great care is needed at this stage to keep the ins and outs as square as possible with the headland mark.

Five Furrows Left To Plough

On completion of this run, careful measuring while ploughing the main part of the plot will have made sure that exactly five or six furrow widths are left to plough and there is enough land left for both tractor wheels to run on unploughed ground. The plough should be set slightly shallower than on the previous run with both bodies turning a full-width furrow.

Reducing the land from five to three furrow widths with the front left wheel on unploughed land.

Moving the plough sideways on the cross shaft to leave exactly one furrow width for the final run of the finish.

Three or Four Furrows Left To Plough

If four furrows are left to plough two must be turned on the next run. With two or three furrows left several adjustments are necessary before reducing the unploughed ground to a single furrow width. Repositioning the depth wheel sideways behind the left tractor wheel will help to keep the plough at an even depth. On ploughs where the depth wheel cannot be moved sideways it may be necessary on this run to regulate depth with the hydraulic lift lever.

The plough needs to be moved across on the cross shaft at this stage so that a single furrow width will be left to plough on the final run of the finish. In World-style conventional ploughing this is done with a screw adjuster on the cross shaft. Without this luxury, after slackening the retaining bolts the plough will have to be moved across manually on the cross shaft.

With the plough re-set on the cross shaft, lower the disc coulter on the rear body so that it runs alongside the share. Lengthening the right lift rod and the top link will help to keep the plough level during this run. Some fine tuning on the lift rod levelling box and depth wheel will probably be needed to get both bodies to turn shallower furrows than those of the previous run with the front body the deeper of the two.

The front left tractor wheel must run alongside the furrow wall and, in common with previous runs of the finish, great care is needed to plough the straightest possible furrow. The front body should turn a two-thirds depth furrow and the rear body should plough a 8–10 cm (3-4 in) deep furrow to leave a single unploughed furrow width for the final run.

It is important to ensure that the rear furrow is not too deep so that enough soil remains for the rear body to turn a sole or earth furrow on the final run.

With three furrows left to plough the front left tractor wheel needs to run close up against the furrow wall.

The Final Run

Ploughing the remaining full furrow width of undisturbed land, together with a sole furrow from the bottom of the furrow left by the back body on the previous run, will complete the plot.

The plough must be returned to its normal position on the cross shaft before ploughing the final furrows. The rear depth wheel should remain behind the left tractor wheel and repositioning the front depth wheel underneath the plough will help to achieve an even depth throughout

The rear depth wheel follows the left tractor wheel while the front body turns a two-thirds depth furrow and the rear body turns a shallow furrow to leave enough soil for the rear body on the final run.

Using the screw adjuster to lower the front disc coulter to run alongside the share for the final run of the finish.

The front right tractor wheel needs to run against the furrow wall on the final run to complete the finish.

the full length of the last furrows. The front body should be raised with the levelling box lift rod so that it ploughs a half-depth furrow while the rear body turns a 8–10 cm (3-4 in) deep sole furrow from the furrow bottom left on the previous run.

On the final run the front right tractor wheel needs to run against the furrow wall with the left wheel following the mark left by the right tractor wheel on the previous run. Points will be lost if more than one tractor wheel mark is visible on the completed finish.

The furrows in this are uniform and there is no trash showing or any loose soil in the bottom of the last furrow. The wheelmark is good and level with the other furrows and the last shallow furrow next to the wheelmark is neat and tidy.

The judges will look for a finish that is uniform throughout its length, shallow and straight. All of the land must be cut with the last eight furrows sloping gently down to a final shallow open furrow. The tractor wheel mark should be level with the furrows ploughed to complete the finish. Points will be lost if two tractor wheel marks are visible on the completed finish.

A good plot with an excellent start ploughed with a World-style conventional plough.

7. Reversible Ploughs

Three hours are allowed for ploughing a plot in the reversible class at a ploughing match.

Competition plots are marked out with a shallow scratch furrow turned towards the headland at both ends of the plot. Competitors are allowed to spread this soil to help them keep the ins and outs neat and tidy. The plots are marked out so that the opening and finishing furrows on each plot are ploughed at an angle to the headland mark.

Reversible match ploughs have more working adjustments than a standard commercial reversible plough.

When the eight or nine furrows required for the start have been ploughed each competitor will have a triangular area with full-length furrows and short work, or butts, to plough before ploughing the last 19 or 20 furrows to complete the plot. An eight-furrow start is used for a two-furrow plough; nine furrows is used for a three-furrow plough.

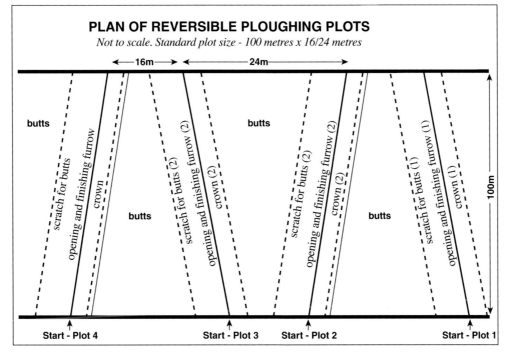

PLAN OF REVERSIBLE PLOUGHING PLOTS

Not to scale. Standard plot size - 100 metres x 16/24 metres

Preparing the Tractor and Plough

When ploughing with a two-furrow reversible plough the rear tractor tyres should be no more than 13 in (32 cm) wide with the wheel centres set at a maximum of 56 in (142 cm). The hydraulic linkage lift rods should be set at the same length before attaching the left lift arm followed by the right lift arm and finally the top link.

Most competitors in reversible ploughing classes use a hydraulic top link. The rules also allow the use of a hydraulic ram to alter the length of the left-hand lift rod and a ram on the plough is used to move it sideways on the tractor.

A hydraulic top link and a hydraulically adjusted left-hand lift rod are used by many competitors in reversible and World-style conventional ploughing classes.

The Eight- or Nine-Furrow Start

The start consists of an opening furrow ploughed with the right rear body which is turned towards the neighbouring plot with the next highest number. After the opening furrow has been judged, eight or nine full-width furrows are ploughed towards the neighbouring plot with the next lowest number.

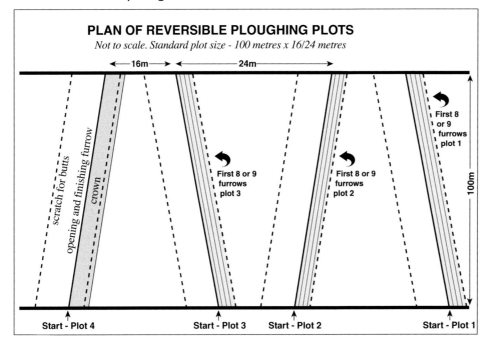

PLAN OF REVERSIBLE PLOUGHING PLOTS

Not to scale. Standard plot size - 100 metres x 16/24 metres

←16m→ ←—24m—→

scratch for butts
opening and finishing furrow
crown

First 8 or 9 furrows plot 3

First 8 or 9 furrows plot 2

First 8 or 9 furrows plot 1

100m

Start - Plot 4   Start - Plot 3   Start - Plot 2   Start - Plot 1

The opening furrows for each plot are set at an angle to the headland scratch mark. This gives a triangular area of land on each plot with full-length furrows and some short work or butts. The start or crown has been ploughed and scratch marks have been made for the butts to leave 19 or 20 furrow widths to plough to complete the finish.

The Opening Furrow

Several adjustments have to be made before ploughing the opening furrow. The right rear disc coulter needs to be lowered to run alongside and below the share. On some ploughs there is a risk of the front share digging into the ground while ploughing the opening furrow. As points may be deducted for this error it is a wise precaution to remove this share.

With the plough prepared for work, lengthen the top link to lift the front right body well clear of the ground, an easy task with a hydraulic top link. Set the plough to turn a shallow furrow 6–8 cm (2-3 in) deep with the rear body. This furrow may need to be a little deeper on light land.

If there is a hydraulic ram on the left lift rod it can be used to shorten the lift rod and raise the front body even higher, further reducing the depth of the furrow turned by the right rear body.

Carefully line up the centre of the tractor bonnet with the sighting poles. If the tractor has a ram on the right lift arm this facility can be used to move the plough sideways to line up the right rear share with the marker peg.

The plough is now ready to start ploughing a 6–8 cm (2-3 in) deep opening furrow with the rear body. Drive forwards a yard or so then stop to check that the opening furrow is at the correct depth. Make any necessary adjustment to the depth of the furrow, then continue ploughing, but only after settling back in exactly the same position on the tractor seat.

The front right body has been raised clear of the ground and the share removed before ploughing the opening furrow.

Complete concentration is essential at this stage and looking back at the work without first stopping the tractor is almost bound to put a bend in the furrow. When leaving the tractor for any reason it is important to take up exactly the same position on the seat before moving off again. The judges will award high marks for a straight, well-cut opening furrow that is uniform throughout its length.

Marking The Butts

Competitors are allowed to plough a very shallow scratch furrow to mark the butts while the judges award their points for the opening furrow. The distance between the scratch mark and the opening furrow must be calculated to ensure that 19 or 20 full furrow widths will be left to plough towards the butts to complete the plot.

Accurate measuring at this stage and careful alignment of the sighting poles are essential to make sure that the correct width of land will be left to plough for the finish. The scratch mark must be straight and parallel with the opening furrow because it is the only driving guide allowed to show when to lift the plough when ploughing the butts and for the first run of the finish.

Ploughing a straight and shallow opening furrow with the right rear body.

The front body needs to be raised well clear of the ground while ploughing a small scratch mark for the butts with the rear body.

Ploughing the First Run of the Eight- or Nine-Furrow Start

On this run the front left body will plough back the shallow opening furrow slice together with some uncut soil from below to give a half-depth furrow while the rear body ploughs a slightly deeper furrow. Both disc coulters should be set a little lower than they would be when ploughing at the minimum depth of 20 cm (8 in) for the reversible ploughing class. Set the rear skimmer a little lower than it would be for normal work and lift the front skimmer so that it is clear of the furrow slice turned on the opening run.

The first two furrows of the eight-furrow start are turned towards the neighbouring plot with the next lowest number, i.e. competitor no. 3 turns the first two furrows towards plot no. 2.

With the plough adjusted for the first two furrows of the start the front left tractor wheel needs to run against the furrow wall of the opening run with both bodies turning full-width furrows. If there is a side ram on the right lift arm this can be used to move the plough sideways so that the front body ploughs a full-width furrow on this run.

The scratch mark for the butts should be straight and parallel with the opening furrow and no more than 3–5 cm (1-2 in) in depth.

Lower the plough with the front share at the headland mark. When using a tractor with a hydraulic top link, shorten it increase the pitch to give a quick entry into work, then lengthen it again when the front body is running at about two thirds of the minimum 20 cm (8 in) competition depth. The front share and disc must run exactly up to the edge of the mark cut by the disc for the opening furrow.

The front left tractor wheel should run against the furrow wall when ploughing the first two furrows of the start.

The top edge of the front furrow slice should line up with the furrow wall from the opening run.

The completed first run of the start with two-thirds depth furrows which are straight along their entire length.

The top edge of the front furrow ploughed on the first run of the start should be exactly in line with the furrow wall made on the opening run. Placing a hand against the furrow wall is a simple way to check the accurate position of this furrow.

The shape of the front furrow slice ploughed on the first run of the start is important. If it is too high or too low this fault will show when ploughing the last furrow for the finish which should be ploughed with the rear disc exactly in line with the edge of the disc cut made for the opening run.

A high, wide, narrow or deep front furrow ploughed at this stage is bound to show up when making the finish, so after driving a short distance, stop the tractor to measure front furrow width and check its shape. If the furrow is too wide or too narrow a slight change of position of the tractor front wheel or moving the plough on the cross shaft should correct the fault.

It is important to stop a couple more times to check front furrow depth and width as turning the correct shape and size furrow at this stage may well prevent problems when ploughing the last furrows of the finish.

Sometimes a plough will turn the correct width front furrow but the furrow slice is still too high. This problem can be overcome by setting the tailpiece lower on the front body to flatten the furrow slice.

As more points will be lost at the finish for an unploughed strip than for over-ploughing it is always better to over-plough rather than under-plough.

Completing the Start

The remaining furrows of the eight- or nine-furrow start must be ploughed at the minimum competition depth of 20 cm (8 in). Before they are ploughed the discs and skimmers need to be put back to their normal settings. If the front tailpiece was lowered for the previous run it should be raised while ploughing the next six or seven furrows. The first stage of ploughing a reversible plot is now complete and the judges with more points to award will look for a level start with even and well-cut furrows.

Ins and Outs

Points are awarded for neat 'ins' and 'outs' so the ends of the furrows must be as neat and accurate as possible, with all the land ploughed up to the headland mark. Always lower and raise the plough into and out of work in exactly the same position every time and make sure the front body takes a full furrow width when it enters the ground.

The remaining six furrows of the eight-furrow start must be ploughed at the minimum competition depth of 20 cm (8 in).

The skimmers on this plough are turning the top corner of the furrow slice into the furrow bottom, leaving no unburied trash.

*Maximum points would probably be awarded for the neat 'ins' (top)
but very few would be awarded for the ragged furrow ends.*

The 'ins' will be even neater if the top link is shortened to pull the front body into work, then lengthened again to put the front and rear bodies at the correct working depth. Competition rules require the plough bodies to be at the specified minimum depth within 1 m (3 ft) of the headland mark.

The ins and outs will be easier to control when using a tractor with a side ram on the right lift arm and a ram on the left lift rod. The lift arm side ram can be used to move the plough sideways to achieve the correct front furrow width at the start of the run. Increasing the length of the left lift rod after ploughing a short distance along the furrow will quickly put the rear body down to its full working depth. The 'outs' can be improved by lifting the front body with the top link while leaving the back body to do a little more work before lifting it almost clear of the ground. Then sweep the end of the furrow slice with the back of the mouldboard as the tractor turns out on to the headland. The left lift rod ram can also be used to raise and tilt the front body a little before using the top link to lift it almost out of work close to the headland mark.

Ploughing the Full-Length Furrows

After the completion of the start the full-length furrows are ploughed at the minimum competition depth of 20 cm (8 in) before ploughing the butts. The first two full-length furrows are ploughed against the start of the neighbouring plot with the next highest number.

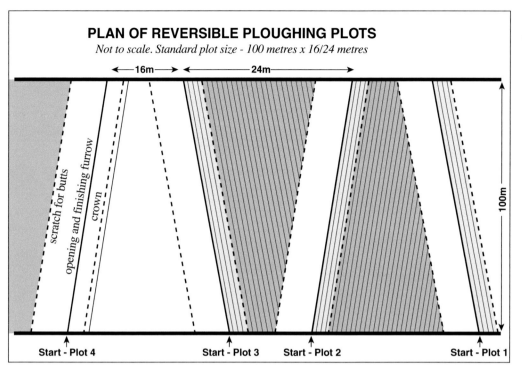

PLAN OF REVERSIBLE PLOUGHING PLOTS

Not to scale. Standard plot size - 100 metres x 16/24 metres

←16m→   ←————24m————→

scratch for butts

opening and finishing furrow

crown

100m

Start - Plot 4   Start - Plot 3   Start - Plot 2   Start - Plot 1

After the judges have awarded their points for the eight- or nine-furrow start the next stage consists of ploughing the full-length furrows and the short work or butts.

Ploughing the full-length furrows always at the minimum required depth, as straight as possible and with no trash showing between the furrows.

Ploughing The Butts

The butts, or short work, must be ploughed at the minimum 20 cm (8 in) depth. With the plough set for normal work, the full-length furrows are ploughed before the butts. Keeping the ins and outs tidy should not be too difficult while ploughing the full-length furrows but it's not so easy where the short work meets the scratch furrow marking the land left for the 19- or 20-furrow finish.

Although the tractor can be turned on the headland at one end of the butts it must be turned on the land left for the finish where the short work meets the scratch furrow. Points will be lost for any wheel marks visible on the furrow slices.

The ins and outs at both ends of the butts need to be as neat as possible and all of the land including the last small triangle at the headland end must be ploughed. The left and right bodies must be used alternately when ploughing the butts, including the last two short furrows.

Scratch Furrow Ins And Outs

The front body will reach the butts scratch mark with two or three feet still left to plough with the rear body. If the plough is taken out of work at this point a hole will be left when the connecting or joining furrow, which forms the first two runs of the finish, is ploughed towards the butts. This problem can be overcome by lifting the front of the plough out of work with the hydraulic top link and leaving the rear body to plough up to the scratch mark.

The tractor must be turned on the 19- or 20-furrow width strip of land when ploughing the butts. Points will be lost if any wheel marks are visible on the furrow slices.

A marker attached to the rear depth wheel is a useful driving guide when ploughing the butts. The rear body is taken out of work when the inner pointer reaches the scratch mark and ploughing continues with the front body until the outer pointer arrives at the scratch mark.

It is much easier to achieve neat ins and outs with a plough equipped with a hydraulic ram to adjust front furrow width. When the pointer for the front body reaches the scratch mark, the plough is lifted from work and moved sideways. After reversing a short distance the plough is lowered to turn the last few feet with the rear body.

There is a similar problem when starting back into work after turning round. With both bodies set to turn equal-depth furrows, lowering the front body back into work at the scratch mark means the rear body will enter the ground too early and start ploughing before it reaches the mark. On the other hand, waiting until the back body reaches the mark before lowering the plough means that ground which should have been turned by the front body will be left unploughed. Both of these faults will be obvious when the judges award their points for the joining furrows.

Without the aid of a hydraulic ram to adjust front furrow width the solution is to shorten the top link to put the front body into work and lengthen it again when the rear body reaches the scratch mark. If the plough has a hydraulic front furrow width adjustment the first few feet of the return run are ploughed with the rear body. After lifting the plough, reversing and moving the front body back to its normal position the plough is lowered back into work. Fine tuning with the top link will ensure both bodies are ploughing at the correct depth.

Ploughing the butts. The plough has been moved sideways to put the front body into the previous furrow while ploughing a single furrow with the rear body. With the inner pointer at the scratch mark it is time to lift the plough out of work.

Ploughing the last two short furrows of the butts. This plough has hydraulic rams on the rear bodies and the right body is being lifted out of work while the front body ploughs up to the scratch mark.

The butts have been ploughed and the next step will be to plough the connecting, or joining, furrow towards the butts.

PLAN OF REVERSIBLE PLOUGHING PLOTS

Not to scale. Standard plot size - 100 metres x 16/24 metres

←—16m—→ ←————24m————→

scratch for butts

opening and finishing furrow

crown

100m

Start - Plot 4   Start - Plot 3   Start - Plot 2   Start - Plot 1

Ploughing the last 19 or 20 furrows will complete the plot.

Ploughing the Finish

The finish consists of 19 furrow widths when using a two-furrow plough or 20 furrow widths for a three-furrow plough. Accurate measuring when setting out the sighting poles for the butts scratch furrow should have made sure that the furrows at both sides of the finish are parallel and there will be exactly 19 or 20 furrow widths between the butts and the first furrow of the start.

Checking these measurements is worthwhile just in case something is not quite right. Any slight error can be corrected while ploughing the joining furrow towards the butts.

The Joining Furrow

As sighting poles are not allowed at this stage of the competition the scratch furrow made earlier is the only permitted driving guide for the front left tractor wheel. On the first run of the finish with a two-furrow plough the front body tidies the ends of the butts while the back body ploughs the joining furrow which is also the first furrow of the 19-furrow finish.

When a three-furrow plough is used the front body tidies the ends of the butts, the middle body ploughs the joining furrow and the back body ploughs the next furrow of the 20-furrow finish.

The position of the front wheel in relation to the scratch mark will be determined by the distance, measured in at least three places, between the scratch mark and the ends of the butt furrows. With all ploughs it is critical to have exactly 18 furrows widths left to plough after the joining furrow has been turned towards the butts on the first run of the finish.

The joining furrow must be at full furrow depth and fully visible throughout the length of the plot. Inaccurate ins and outs while ploughing the butts will show up as holes and mounds between the ends of the butt furrows and the furrow turned by the front body.

The scratch mark is the only driving guide allowed while ploughing the joining, or connecting, furrow.

Ploughing the joining furrow. The judges have more points to award for the joining furrow and points will be lost if the front furrow is not fully visible and straight throughout its length. Still more points will be lost if inaccurate butt ends result in the appearance of an odd hole or mound between them and the joining furrow.

Care with the ins and outs while ploughing the butts has resulted in a joining furrow with no obvious holes or mounds between it and the ends of the butt furrows.

Inaccurate ins and outs result in holes or mounds and wheel marks between the joining furrow and the butts.

Measuring the distance between the rear disc coulter and the edge of the shallow opening furrow to check that they are parallel and that there are exactly 18 furrow widths left to plough while ploughing the joining furrow. The rear body will need to line up with the edge of the opening furrow when the last two furrows are ploughed to complete the finish.

When the joining furrow has been ploughed it is important to measure the distance between the two sides of the unploughed strip in at least three places along the length of the plot. If the two sides are not quite parallel the error can be corrected by ploughing a slightly wider or narrower furrow where necessary on the next run. With a little luck the correction will not be noticed by the judges.

Completing The Finish

Normal ploughing continues until the unploughed land is reduced to four furrow widths when using a two-furrow plough or six furrow widths with a three-furrow plough. With three hours allowed for ploughing the plot there should be plenty of time to check once again that the two furrow walls are still the correct distance apart and parallel along their length.

When using a three-furrow plough the unploughed land will be reduced to six furrows leaving three furrow widths for the final run.

When the unploughed land has been reduced to four furrow widths with a two-furrow plough the front right tractor wheel should run alongside the start furrow while ploughing the next two furrows. This will leave two furrow widths for the final run of the finish.

The rear depth wheel on the plough needs to run alongside the first furrow slice of the start while reducing the unploughed land to the two full furrow widths.

The Final Run

Careful measuring when setting out the scratch furrow for the butts and while ploughing earlier furrows for the finish should have made sure that exactly two furrow widths are left to plough on the final run with a two-furrow plough or three full furrow widths when using a three-furrow plough.

More points are allocated for accuracy in ploughing the last furrow of the reversible finish. The judges will look for a furrow wall ploughed close up against the furrow slice that was turned on the opening run. Points will be lost if the rear disc hub rubs soil off the opening furrow slice and leaves loose soil in the furrow bottom. Replacing the normal disc coulter with one which has a right-handed hub that will run above the unploughed side of the disc is allowed and will overcome this problem.

Before ploughing the last two or three furrows to complete the finish the rear depth wheel should be moved sideways to make it follow the rear right tractor wheel.

Points will be lost if the rear disc coulter hub rubs soil off the opening furrow slice. Reversing the disc coulter assembly so that the hub runs above unploughed land avoids the problem.

The front left wheel needs to run alongside the furrow wall while ploughing the last two furrows of the finish.

The rear body should run exactly at the edge of the disc cut made when the opening furrow was ploughed with the plough depth wheel following in the wheel mark made by the rear right tractor wheel.

An example of a well-ploughed plot with straight, uniform furrows with no hills, holes or wheel marks between the butts and the joining furrow.

A good reversible finish, which is straight and level with all the soil ploughed up to the disc cut of the opening run. The wheelmark is level with the uniform furrows.

This finish has faults that the judges are bound to see. The last furrows are very untidy and uneven and have not been cut to the disc cut made on the opening run leaving an unploughed strip of land.

8. High Cut or Oat Seed Furrow Ploughing

High cut or high crested ploughing, also known as oat seed furrow, is a type of ploughing upon which cereal seed can be sown by hand. The furrows are narrow and tightly packed together in the shape of equilateral triangles with a sharp comb so that after the seed has been sown it can be covered with a single pass of the spike-tooth harrows. The high angle of the furrows promotes rapid drying of the soil so that after the seed has been broadcast the passage of the harrows easily crumbles away the tops of the furrows to cover the seed.

Oats were, and usually still are, grown in 7 in (18 cm) rows. For this reason furrow widths of 15–18 cm (6–7 in) with a minimum depth of 13 cm (5 in) are specified for the oat seed furrow or high cut classes at ploughing matches.

The high cut plough has a long mouldboard with a gentle convex curve. The wing of the share is raised and the disc coulter set at an angle to turn a furrow with a sharp comb. This is known as the cut, and the sharper the angle of the cutting face of the coulter to the cutting face of the share the higher the comb or cut to the furrow. The other main difference between a plough used in the trailed vintage class and the high

Competitors are allowed to use various furrow-shaping aids in the oat seed furrow class at ploughing matches.

cut class is the permitted procession of boats, seamers, press wheels and chains pulled behind the bodies. These shape the furrow slices and pack them tightly together to prevent the hand-broadcast seed from falling through between the furrow slices. The tightly packed furrows also prevent the emergence of weeds that might compete with the seedling oat crop.

Ploughing a plot at a ploughing match with press wheels and boats to shape the cut of the furrows in the high cut or oat seed furrow class.

Competitors have five hours to complete a plot in the oat seed furrow class at a ploughing match. Points are allocated for the start, for the ins and outs, for straight and uniform furrows and for the finish. The judges have further points to award for firm and well-packed furrows together with the quality of the seedbed and the general appearance of the completed plot.

An oat seed furrow plot is ploughed in a different way from a plot for a vintage trailed plough. Sighting poles are used to set out the first two furrows of the start. There is no opening split and the start consists of 11 or 12 furrows. The minimum required ploughing depth of 13 cm (5 in) must be reached by the completion of the start and must be maintained until ploughing the eight furrows that form the finish.

A top quality high cut plot ploughed by veteran ploughman Mervyn Vowles who competed in over fifty British Ploughing Championships.

9. A Short History of the Plough

A Saxon wheeled plough.

Early Ploughs

In his book *The Implements of Agriculture* originally published in 1843 and reprinted by Old Pond Publishing in 2003, James Allen Ransome wrote 'The plough is certainly the most valuable and the most extensively employed of all agricultural implements.'

References to this implement can be found in the Old Testament and some 1,900 years ago the Romans were using the ard, a crude form of wooden plough, pulled by a pair of oxen. Early ploughmen also used camels, asses and elephants as draught animals and it was not unknown for a plough to be drawn by the farmer's wife and steered by the farmer himself. As time passed peasant farmers fashioned tree trunks into shapes resembling a plough beam to carry a very basic wooden mouldboard and a handle to guide the plough. (The term mouldboard is derived from the use of a wooden board to invert the soil. In earlier days mouldboard was a mainly American term with breast being the more usual name in Britain.) The primitive breast plough was, as the name

suggests, pushed by the farmer wearing a protective shoulder apron.

The Romans were making wrought-iron shares in pre-Christian days and 600 years later the Saxons developed a wheeled horse- or oxen-drawn plough which survived beyond the days of the Norman Conquest. They brought their ploughs with a wooden beam, mouldboard, share, simple coulter and wheels to Britain in the sixth century at a time when some farmers may have been yoking their horses by their tails to a plough. This practice continued in parts of Ireland until 1634 when the Irish Parliament found it necessary to pass a bill forbidding the 'barbarous custome of ploughing with horses by the taile'. An early British law required that any aspiring ploughman should make his own plough and not be allowed to guide a plough in the field until he had completed this task.

Roman ploughmen turned furrows about 40 yards (40 m) long with a plough drawn by two oxen but the Anglo-Saxons with eight oxen

GENERAL PURPOSE WOOD BEAM PLOUGH.

Wood Plough, W.R.N.E., with long match breast.

This Plough is adapted for light or medium soils, and will work easily with two horses; it is very simple and strong in its construction, and will produce the best work, as it can be fitted with the same shares and breasts as the celebrated Newcastle and Leicester Prize Series of Ploughs.

The Breasts and Shares for the W.R.N.E. Plough are marked **R.N.F.**

W.R.N.E., with Round Coulter, Two wheels, **£4 2s.** One wheel, **£3 14s.** Swing, **£3 4s.** Skim Coulter, 5/- Steel instead of Cast Breast, 5/- Drag Weight and Chain, 2/- extra.

A Ransomes, Sims & Head wood beam plough from their 1884 catalogue.

ploughed furrows 220 yards (220 m) long. Known as a 'furrowlong' (later furlong), it was considered that 220 yards was the greatest distance oxen could walk without stopping for a break. A Saxon farmer with his team of oxen was able to plough one acre, or a strip of land 220 yards long and 22 yards wide, in a day. Even in relatively modern times the opening furrows would be set 22 yards apart for a two-furrow tractor plough and at 33 yards for a three-furrow plough.

In the late seventeenth century the Dutch were improving their wooden ploughs with the addition of a wrought-iron facing on the mouldboard. Some of these ploughs were used by Dutch drainage engineers in the late 1720s to help reclaim fenland in Lincolnshire. A Yorkshire engineer called Joseph Foljambe patented an improved version of the Dutch plough in 1730. Known as the Rotherham plough, it was a wooden-framed swing plough with no depth wheels; the wooden mouldboard and share were clad with wrought iron and it had an iron coulter.

Around this time farmers and engineers turned their attention to developing bigger and better ploughs. Some built them with either one or two wheels mounted on a cross frame at right angles to the main beam. The cheaper swing plough was said to be an advantage on sticky soils where wheels would soon become clogged. However, as a writer of the day complained, the swing plough required constant attention from the ploughman who would soon suffer from fatigue and whose output would be reduced. The gallows plough, recommended for deep work, was popular in East Anglia. The wheels were carried on a separate fore-carriage, which also supported the beam, and the plough was steered with two chains linked to the fore-carriage.

Scotsman John Small applied both mathematics and science to the shape of the mouldboard. He became the first to use a cast-iron mouldboard designed to turn the furrow naturally rather than tearing the furrow slice away. Used on his iron-framed Scotch or Berwickshire plough, made from about 1775, the new cast-iron mouldboard was easier to pull, suffered less wear and turned the soil more effectively.

In some countries the more superstitious members of the farming community believed that pulling iron through the soil would poison it. This superstition still survived among American farmers when the first iron plough, designed by Charles Newbold, appeared there in 1797. Plough shares were all made of wrought iron until the mid-1770s when, among others, Robert Ransome started making cast-iron shares at his Norwich foundry. The next landmark came in 1785 when Ransome was granted a patent for his method of making and tempering cast-iron shares with salt water. He took out a further patent for chilled cast-iron plough shares in 1803, this time from his new foundry at St. Margaret's Ditches in Ipswich. The process consisted of chilling the under-surface of the share to make it extremely hard, leaving the upper surface soft and tough so that it wore away more quickly to keep the share sharp.

Another Ransome patent, granted in 1808, was destined to become an important feature of both agricultural and general engineering design. The patent was initially concerned with the manufacture of plough bodies in a way that made it possible to remove worn or damaged parts and replace them with identical new parts.

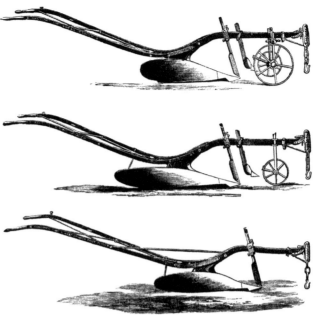

Double-wheeled, single-wheeled and swing-type Newcastle Series Ransomes ploughs.

In the same year a Mr. Simpson, a farmer at Cretingham in Suffolk, added a slade to the bottom of the plough body to take the downward thrust created by the weight of the plough and the furrow slice on the mouldboard. The very fact that ploughs were being made by local blacksmiths meant that there were many variations of design.

The Royal Agricultural Society of England held a series of ploughing trials during the mid-1800s. The Southampton Trial of 1844 awarded first prize to the Ransome YL (Yorkshire Light) plough. Originally with a wooden beam and later as an iron plough the Yorkshire Light was made for at least 100 years. The Newcastle Trials of 1864, where Ransomes swept the board with their single-furrow iron ploughs, were a particular landmark in plough history. The judges were looking for ploughs that 'cut the sole of the furrow perfectly flat, lay the furrow slices with

A Howards of Bedford gallows plough.

uniformity and perpendicular cut on the land side, leaving a roomy horse walk'. The winning Ransomes ploughs, which became known as the Newcastle series and supplied with one or two wheels or as a wheel-less swing plough, were made at Ipswich for the next 80 years.

Ploughs were also made in vast numbers at the Britannia Iron Works at Bedford. Founded in 1813 by John Howard, he made his first iron plough with wheels in 1839 and plough manufacture continued there for almost 100 years. The 1839 iron plough was the forerunner of the Howard Champion plough destined to be made in a steadily improved form for around 80 years. By 1879 and now trading as James and Fredk. Howard Ltd, their farm implements catalogue listed 12 different types of plough, some with wooden beams for both home and export markets.

Howards needed to have no doubt about the quality of their products. In their 1879 catalogue they printed a list of prizes that included 56 awards from the RASE including a special prize for the Howard self-acting horse rake and a gold medal for steam cultivating machinery. The catalogue also pointed out that for several years J & F Howard had not competed at any of the so-called All-England ploughing matches believing that instead of sending 'highly trained men and horses with ploughs specially got up for such matches' it was a fairer criterion to leave their ploughs entirely in the hands of local ploughmen. There may have been a touch of sour grapes about this decision as Ransomes had publicised the fact that their employee and champion ploughman James Barker had won over £2,000 on his travels around the country with a pair of Suffolk Punches and a Newcastle plough.

John Pendry of Cornwall ploughing his way to victory at the 1955 British Ploughing Championships in Flintshire.

At the end of the First World War J & F Howard Ltd. joined forces with 13 other companies to set up the Agricultural & General Engineers Association (A&GE). The plan was for member companies to sell each other's products and the Howard of Bedford catalogue for 1924 also included Aveling & Porter, Charles Burrell, Richard Garrett and Davey Paxman steam engines and Peter Brotherhood's Peterbro' tractor. Unfortunately for James & Fredk Howard, the A&GE went into liquidation in 1932 and the agricultural machinery side of their business was acquired by Ransomes, Sims & Jefferies.

British-made ploughs became famous worldwide, perhaps largely because of the number of ploughs taken by emigrants in the 1840s and 1850s to use on their new lands in the colonies. Ransomes' catalogue for 1851 noted that their ploughs had been sent to America, South Africa, Australia and the West Indies and that special designs, including elephant-draught ploughs for use on Indian sugar plantations, had also been made.

Ploughs were also being made at this time in blacksmiths' shops across America. William Parlin, one of many farm machinery pioneers in America, made his first plough in his Illinois blacksmith's shop in 1842. Ten years later he formed a partnership with William Orendorff and a range of P & O ploughs soon became popular with American farmers. The International Harvester Company eventually acquired the P & O line of farm implements in 1919.

Scotsman James Oliver, who gave his name to the American Oliver Corporation, was another pioneering ploughmaker. Having emigrated with his parents to America in 1834 aged 11 he

bought a share in a local iron foundry. After several years of experimentation he perfected a chilled-iron mouldboard claimed to give a reduction of 20–50% in the draught of his ploughs. One hundred years later the Oliver Corporation were making more ploughs than any other company in America.

The growing popularity of multi-furrow steam-powered balance ploughs in the latter half of the nineteenth century resulted in a demand for horse ploughs with two, and in some areas, three bodies. Two-furrow ploughs had originally appeared in the mid-1600s, hauled by large numbers of oxen or horses. The extremely wide headlands needed to turn these outfits made them totally impracticable.

Howards of Bedford, Ransomes and several other ploughmakers had met these demands by the late 1870s with both wooden- and iron-framed two-furrow ploughs. Light land farmers were even able to buy a three-furrow plough which, in some cases, could be converted for two-furrow work when the land was too wet for three. One farmer near Ipswich was reported to have three two-furrow ploughs, each pulled by three horses, so that each ploughman was able to plough two acres in

A high-beam animal-draught plough for two oxen which dates back to the late 1800s.

A Ransomes horse plough with the Jefferies patent lift.

Although rich estate owners turned to steam ploughing in the latter half of the nineteenth century, most farmers still ploughed with horses. American farmers made life easier by using single furrow ride-on, or sulky, ploughs a design taken up by Ransomes and others who exported large numbers to Argentina.

a day rather than the hitherto single acre of land. Although a number of these ploughs were sold in Britain they were far more popular with American and Canadian farmers.

Handling a two- or three-horse plough was hard work, especially on heavy soils, but life was made easier for the ploughman when Mr. J. R. Jefferies patented the Jefferies lift in 1870. This double-wheel lifting apparatus, consisting of a lever connected to a pair of wheels on a separate cranked axle, was designed to lift or lower the plough at the headlands and set ploughing depth.

Cast-iron plough shares were prone to damage on shallow soils with rocky outcrops so some ploughmakers offered their customers the alternative bar-point body in the late 1920s. The body was specially designed for use with a long chisel-pointed steel bar in place of the more usual share that could be moved forward to compensate for wear. Some makers used a bar pointed at both ends to double its working life. A later design of bar-point body was spring-loaded, allowing the point to recoil on striking a solid object and then rebound forward, sometimes shattering the obstruction (see page 10).

Pulled by two or three horses, the sulky had an automatic power-lift operated by the wheels and the ploughman was able to vary working depth from his seat.

A single-furrow Ransomes sulky, or riding, plough with an automatic power-lift.

Early One-Way Ploughs

The technique of one-way or reversible ploughing, using a plough with one or more pairs of left- and right-handed bodies, pulled back and forth from one side of a field to the other has been practised for well over 150 years. Horse-drawn turnwrest ploughs date back to the 1840s and balance ploughs also pulled by horses appeared in the 1860s. The Davey-Sleep single-furrow balance plough, made in south-west England, is an early example of a horse-drawn balance plough. Several ploughmakers copied the Davey-Sleep design and multi-furrow balance ploughs were a common sight during the steam age. The French Brabant turnover plough for two or more horses with the idle body or bodies at 180 degrees to those in work were being made in Britain and continental Europe by the mid-1800s.

The body of Kent turnwrest plough, one of the earliest horse-drawn reversible or one-way ploughs, had a central share in front of a short split mouldboard with left and right wings similar to a ridging body. A detachable wooden 'ground wrest' or mouldboard extension was alternately pegged to the left and right mouldboard wings. On reaching the headland the 'ground rest' was transferred to the opposite mouldboard wing for the return run across the field. Ransomes' catalogue for 1886 explained that the movement of a lever on their SPT turnwrest plough 'turned the share over, and while one breast is put into its proper position for work, the other is raised and carried clear on the land side'. The body on later types of turnwrest plough was swung from one side of the beam to the other while turning at the headland for the next run.

The horse-drawn balance plough was a scaled-down version of the steam-powered balance plough first made in the 1850s. It had left and right bodies complete with coulter and steerage handle pivoted on a central axle and carried on two wheels. In use, the plough was tipped to put the opposite bodies into work and the horse had to be moved to the opposite end of the plough before the next furrow could be ploughed. The Lowcock turnwrest plough, awarded a silver medal at the RASE meeting at Southampton in 1844, was described in Ransomes & Sims' catalogue for 1862 as 'simple in its formation and almost self acting, as respects its adaptation to each successive furrow'. It was explained that on arrival at the headland the ploughman 'turns the handle from one end of the beam to the other while the horses turn round on the land side of the plough'. The Lowcock plough could turn a furrow 7 in x 10½ in (18 x 26 cm) and was strong enough for four horses.

Unlike the bodies on a balance plough, which pivoted in the line of travel, the left and right bodies on the Brabant turnwrest were pivoted about the main beam at 90 degrees to the furrows with one body in work and the other vertically above in the air. The Brabant plough was more popular with farmers in many parts of France and Belgium than in Britain where, among others, it was made by Ransomes of Ipswich and by Howards of Bedford who included it in their catalogue for 1879 priced at '£9.0.0 with steel breasts'.

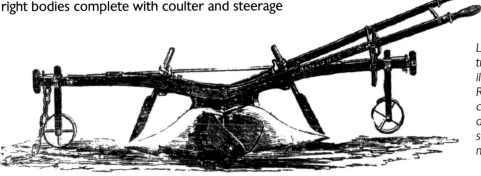

Lowcock's patent turnwrest plough, illustrated in Ransomes' 1862 catalogue, was awarded the first of several RASE silver medals in 1844.

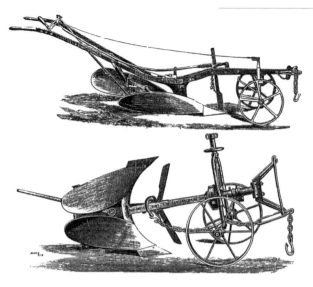

Top – Ransomes' SPT patent turnwrest plough.
Bottom – Brabant turnwrest plough.

Steam Ploughing

Steam power arrived on the farming scene in the 1850s. James Watt took out a patent for a system of steam ploughing in 1770 and a Mr. Heathcote was using steam power in 1832 in an attempt to reclaim some land on his farm at Tiverton in Devon. The next 35 years saw almost 100 provisional patents registered for steam cultivation. The most important was taken out by John Fowler in 1851, for his system of cable mole ploughing, but several years passed by before it was used to any significant extent.

The first public showing of steam ploughing was probably in Essex in 1848, when a system of roundabout cable ploughing, later taken up by Howards of Bedford, was demonstrated with a portable engine and a balance plough. An endless rope running round the field was carried on an engine-mounted winding drum and supported by a series of anchor pulleys placed at each corner of the field. One end of the rope was attached to each end of the plough and as the winch pulled the rope round the field the plough was hauled back and forth between the two headlands. Having previously patented his ideas on steam

cultivations and assisted by William Worby, Works Manager at Ransomes of Ipswich, and Essex farmer David Greig, John Fowler designed an alternative system of cable ploughing. His idea was to have a portable engine with a cable-winding drum at one side of the field and a movable anchor pulley on the opposite headland with the two ends of the cable attached to a balance plough. The engine was to pull the plough back and forth across the field and as work progressed the engine and the anchor pulley would be moved along the headlands.

In 1856 John Fowler commissioned Ransomes to build a balance plough and related tackle. Following successful field trials using a Ransomes portable engine on the nearby Nacton Heath, Fowler's new system of steam ploughing was introduced to the farming public at the 1856 Chelmsford Show. It was noted that the amount of land ploughed in an hour with steam power would take all day to complete with a team of horses. Not satisfied with his success, John Fowler then developed an even more efficient system of cable ploughing with a pair of engines equipped with winding drums that were used to pull a four-furrow balance plough from one side of a field to the other and then back again.

Ransomes manufactured steam-ploughing tackle for John Fowler at the Orwell Works in Ipswich until 1862 when he opened his own factory in Leeds. J. & H. McLaren Ltd, another famous steam engine maker in Leeds, adopted the double engine system of cable ploughing and built their own designs of ploughing engines and balance ploughs.

The two sides of balance ploughs were designed with both halves of equal weight to help the ploughman bring the opposite set of bodies down into work at the start of the next run. However, there was a tendency for the side of the plough in the air to cause the working side of the plough to ride up and not plough as deep as it should. This problem was solved in 1885 when John Fowler & Co invented their anti-balance plough. Its redesigned undercarriage allowed the pull of the cable to move the axle forward beyond the point of balance, making the working half the heavier part so that the shares remained at their required depth.

Steam-powered cable ploughing was not a particularly common practice in the early 1900s when most of Britain's farmland was still being ploughed with horses at a rate of an acre a day with a single-furrow plough. Proportionally more land was ploughed with a two- or three-furrow plough, chain-hitched to a team of horses and, more often that not, a seat at the back for the steersman.

In a last-gasp attempt to keep steam ploughing alive a few farmers hitched a plough by a single

RANSOMES & SIMS' CATALOGUE, 1859.

FOWLER'S STEAM PLOUGH.

ENGINE AND WINDLASS.

PLOUGH.

ANCHOR.

Fowler's ploughing tackle for steam ploughing with a portable engine.

John Fowler's double-engine steam cable ploughing system.

chain to the back of a steam tractor. The plough was a modified version of the two-furrow and three-furrow chain-drawn horse plough. However, as it was not possible for the engine driver to control the plough, a seat was provided for a steersman who was required to keep the plough on course and to put the bodies into and out of work at the headlands. Direct traction steam ploughing enjoyed a degree

A Ransomes steam tractor at work with a RYLT tractor plough.

of success in some parts of the world, but most British farmers preferred to hitch their plough to one of the new breed of farm tractors which, although built to look like a steam engine, had an oil engine under the bonnet.

Ploughing
With Electricity

A patent was issued to a Mr. H. Cooper and John Fowler in 1919 for a cable winding drum centrally mounted under a lookalike traction engine with an internal combustion engine driving an electric generator which in turn supplied current to an electric motor. The motor was to be used to drive the winding drum or the rear wheels. By stationing one of these units on opposite headlands a field could be cable ploughed or cultivated in the same way as by a pair of steam-driven ploughing engines. However, in the event, little more was heard of this particular patent.

Cable-operated balance ploughing using powerful electric motors made more impact on the farming scene than John Fowler's electric ploughing engine. In 1924, agricultural engineers in France and Italy introduced similar methods of cable ploughing using electric motors supplied with current from high-tension overhead power lines. One such system, shown by The Société d'Electro-Motoculture at a Paris exhibition in 1930, consisted of two electric motor-driven portable cable winding drums together with transformers on wheels, cable-drawn balance plough and living van for the plough team. The transformers were used to convert the local 15,000 volt overhead mains supply down to 500 volts. This was connected to a 45 hp winding drum motor on

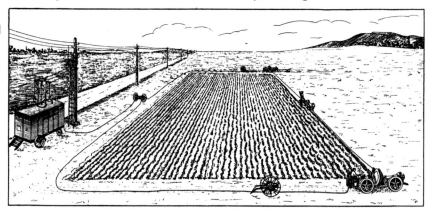

Arrangement of the tackle used for electric cable ploughing.

both carriages and to 10 hp motors used to propel them along the headland. However, oxen were required to move the balance plough, cable winding drum carriage and transformers from field to field.

At the same time, Italian engineers Fabbrica Arati Meccanici were promoting their round and round cable ploughing system with a balance plough, portable transformer linked to the overhead mains supply and an electric motor driving the cable winch. Another Italian company, Camillo Sacerdoti of Milan, exhibited an electric tractor for direct ploughing which automatically took in and paid out an electric cable as it proceeded across the field.

Early Tractor Ploughs

Internal combustion-engined farm tractors were gradually replacing steam power by the early 1900s although some were almost as heavy as a traction engine. More progressive farmers though it better to equip a pair of these tractors with cable winding drums, stand them on opposite headlands and plough their land with a multi-furrow balance plough. The Fowler oil-engined cable ploughing tractor introduced in 1912 had a horizontal cable-winding drum mounted under the fuel and cooling water tanks disguised to look like a steam boiler with a false chimney stack adding to the deception.

four-stroke power unit. The Victoria had two forward and one reverse gears with a top road speed of 5 mph (8 kph). The single-speed cable drum held up to 450 yards (450 m) of ½ in (12 mm) steel rope. A pair of engines could plough between seven and ten acres in a ten-hour day. From a distance the Victoria, with its boiler-shaped tanks for fuel and water and an exhaust pipe with the dimensions of a steam engine chimney stack, could easily be mistaken for a steam engine. The last pair of Victoria ploughing engines was made in 1922.

The Ivel and other models of lightweight tractor made precious little impact on the farming scene during the early years of the twentieth century when horses reigned supreme. The move towards tractor power was given a slight boost in 1907 when the first Model F Fordson tractor arrived from America. Although the UK tractor population had increased to around 15,000 by 1920, including 7,000 Model Fs from the Ford factory at Dearborn, they were still hopelessly outnumbered by the 962,000 working horses on Britain's farms.

The first chain-pulled tractor ploughs appeared in the early 1900s but until 1918 most farmers lucky enough to own a tractor were still using a

The Walsh & Clarke Victoria ploughing engine, probably the best known of its type, was awarded a RASE silver medal in 1915. Victoria ploughing engines had a 45 hp twin-cylinder horizontally opposed paraffin engine; a few had a two-stroke engine but most had a

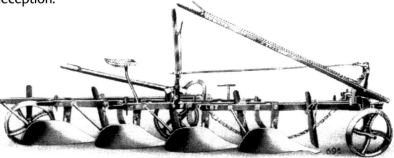

The LBZ tractor plough, made by Howards of Bedford, could be converted to three furrows when necessary. To avoid confusion when ordering farm implements it was common practice for manufacturers to allocate a code word to every model of implement; ELBEZED was the code for the Howard LBZ tractor plough.

two- or three-furrow horse plough chain-hitched to the drawbar. It was usual for a seat to be provided on the plough for a second person to steer the plough and put it into and out of work at the headlands. Ransomes introduced the steerable two-furrow YLTM (Yorkshire Light Tractor Multiple) trailed tractor plough in 1909 and by 1914 had added three- and four-furrow trailed models. The LBZ tractor plough made by J & F Howard at Bedford, introduced at much the same time, was a four-furrow model with a seat for the ploughman. Weighing 12 cwt it could be supplied with knife or disc coulters.

The introduction of the RSLD – Ransomes Self-Lift Double – in 1919 meant that the driver was able to put the plough into and out of work from the tractor seat.

Even though ploughing up the land for wartime food production was a top priority, plough manufacture was somewhat restricted during the war years. It was not until hostilities were at an end that the self-lift tractor plough, giving the tractor driver full control of his plough, arrived on the scene. Ransomes were first in the field in 1919 when they introduced the two- and three-furrow RSLD and RSLM ploughs with a cord-operated self-lift mechanism. The plough was lifted from work by pulling on a cord to engage a curved toothed bar with a sprocket on the landside wheel hub. This caused the plough axles to rotate in their housings and in so doing lift the bodies clear of the ground. A second pull on the rope released the lifting mechanism and the bodies dropped back into work. The depth and front furrow width control levers, previously operated by the steersman riding on the plough, were turned round to put them within reach of the tractor driver. Screw handles replaced the hand

levers on the RSLD and RSLM No 2. Other ploughmakers soon followed suit with their own designs of self-lift plough.

Farmers had a wide choice of trailed tractor plough during the good and bad times between the two world wars, with many made in Britain and others imported from continental Europe and America. Although Ransomes, Sims & Jefferies enjoyed a major share of the UK market other British companies including Hornsby, Howard, Fisher Humphries and Geo. Sellar sold considerable numbers of tractor ploughs in the 1920s and 1930s. Ransomes also faced competition from several American manufacturers such as Cockshutt, International Harvester, Massey Harris and Oliver.

Motor Ploughs

Several designs of motor plough appeared during the early years of the twentieth century. Unlike tractor-drawn ploughs, they did not require a second person to steer them. The engine and driving wheels at the front end replaced the horses and the driver seated at the rear was in full control of the plough and his 'mechanical horse'.

The Crawley brothers at Saffron Walden built the first Crawley Agrimotor for use on their Essex farm in 1908. Awarded a gold medal at the 1919 RASE ploughing trials, the 30 hp Agrimotor with a seat carried on a small castor wheel behind the plough for the driver-cum-ploughman was one of the more successful early designs of motor plough. The Crawley, weighing a little less than two tons, was made from 1914, first by Garretts of Leiston and later by the Crawley brothers themselves at Saffron Walden. The last Agrimotor was built in 1927.

The Wyles Motor Plough, made in various forms between 1910 and 1921 was initially a one-furrow machine with a single-cylinder 4 hp petrol engine, steered by the ploughman walking behind in the same fashion as he had done with a horse-drawn plough. Albert Wyles decided more power was required so later models had an 11 hp Lister single-cylinder water-cooled engine, a single forward gear, steel cleated wheels and two Ransomes RBYD plough bodies suspended from the handlebars. A 14 hp Wyles

motor plough made in 1916 was soon taken up by John Fowler at Leeds who also made a ride-on version known as the Fowler Motor plough.

Fowlers also introduced the Rein Control motor plough in 1923. Designed in Australia it had a two-wheeled front power unit with a two- or three-furrow riding plough hitched to a drawbar. The driver controlled the tractor with reins in a similar way to handling a team of horses. Not only could he control the plough from the driving seat but also, by using different tugs on the reins, he could stop and start the tractor, change gear and steer the machine. The Rein Control motor plough was awarded an RASE Silver Medal in 1924 – and then quietly died!

The single speed American Moline Motor Plow appeared in the UK in 1916 and for a while was as well known as the Crawley Agrimotor. A twin-cylinder engine drove the two front wheels and the driver's seat was carried on two smaller wheels at the rear. The plough was suspended from the main frame in full view of the driver.

Ransomes Boon Motor plough.

Introduced in 1927, the Ransomes Weetrac was designed for tractors equipped with a mechanical lift linkage.

Moracre, the single-furrow Deepacre digger and the Bracre for breaking virgin soil and land reclamation. The four-furrow, self-lift Moracre plough, convertible to three furrows when the going was tough, was recommended for crawler and half-track tractors. Described in sales literature as a plough 'which will give pleasure to any good judge of ploughing' it turned furrows 12 in (30 cm) wide and 11 in (27.5 cm) deep. International Harvester were making the McCormick 8A tractor plough for the Farmall M, Ransomes were offering a huge range of trailed ploughs for home and export markets and the Ford Motor Co were building two- and three-furrow Ford Elite ploughs at Leamington Spa.

Originally designed and built by the Eagle Engineering Company at Warwick, the 52 cwt Boon Motor Plough, also made by Ransomes, appeared at the Aisthorpe trials in 1920. The twin-cylinder motor plough had a three-furrow Ransome plough suspended from the handlebars but, unlike the Crawley Agrimotor, Moline, Wyles and similar motor ploughs, the Ransomes Boon seems to have disappeared without trace.

Mounted Ploughs

Harry Ferguson introduced a mounted plough with a mechanical lift for the Fordson Model F in 1919 and the mounted Ransomes Weetrac with a hand lift followed in 1927. The Ferguson Brown tractor with its draft control hydraulic system and fully mounted plough appeared in 1936.

The demand for trailed tractor ploughs remained for another ten years or so but the range of makes and models diminished as time passed. Fisher Humphries, International Harvester and Ransomes were among the leading trailed ploughmakers in the mid-1900s. The Fisher Humphries range included the

The Ferguson general-purpose three-furrow mounted Ferguson plough with hydraulic depth control cost £76 in 1954. The handle for adjusting the front-furrow width was an optional extra.

Harry Ferguson's mounted plough with his patented tractor hydraulic depth control system was way ahead of is rivals in the early 1950s when many of his competitors, including David Brown, Nuffield, International Harvester, Ford and Allis-Chalmers were restricted to using a wheel to control the depth of ploughing. The idea of starting at one side of a field and ploughing to the opposite with a one-way, two-way or reversible plough, as in the days of steam-powered balance ploughs, was also coming back into fashion at this time. However, with the added advantage of the hydraulic three-point linkage, the technique was now here to stay.

Roger Dowdeswell made a reversible plough for his crawler tractor in the late 1960s. Within a few years he was building a wide range of reversible ploughs fitted with Ransomes bodies.

Except for the Ferguson 'butterfly'-mounted plough, models relied to a greater or lesser extent on muscle power to put the opposite set of bodies into work before starting the next run.

As time passed mechanical turnover mechanisms and hydraulic rams took over from manual effort and reversible ploughs with four, five and occasionally six furrows came into widespread use. However, for a brief period in the mid-1970s the ploughmakers lacked the technical ability to build reversible ploughs big enough to use with many of the 200-plus horse power tractors brought in from North America. The only way to make economic use of all this power was to revert back to the conventional right-handed plough with seven or eight furrows. Stubble ploughs, with up to ten bodies for ploughing about 6 in (15 cm) deep at high speed, were already popular on the Continent and now found favour with British farmers, if only for a short period.

Roger Dowdeswell, who had been making reversible ploughs since the late 1960s was one of a number of ploughmakers who came up with the idea of multi-furrow semi-mounted reversible ploughs for the high-powered tractors from North America.

Push-pull ploughs, usually with seven, eight or nine bodies split between front and rear, came to the fore in the early 1980s. It seemed that mounting a plough at the front of the tractor rather than using several front weights to counterbalance the increasingly heavy multi-furrow mounted ploughs was the way forward. Fiskars, Lemken, Rabewerk and Ransomes were among a number of ploughmakers who moved into this new market that reached its peak around 1985. However, difficulties in matching the work done by the front and rear bodies and to a lesser degree the problem of manoeuvring the tractor and its ploughs in confined spaces resulted in a fairly rapid decline in their use.

Following the semi-mounted design, reversible ploughs gradually became wider and longer, culminating in acre-eating articulated ploughs turning 12 furrows in a single pass. Most ploughs have a hydraulic furrow width adjuster operated by the tractor hydraulic system but some older ploughs have a mechanical furrow width adjustment. Although it is still necessary to turn the tractor round at the headland some tractor

Push-pull ploughs were popular for a while in the mid 1980s.

drivers now have the luxury of a computerised headland management system in the cab. This will carry out the full sequence of events required to take the plough out of work, reverse the bodies and sequentially lower the front and then the back of the plough into work.

It would take several days to do as much work with a horse plough as a multi-furrow articulated reversible plough can do in an hour.

10. Rules for Ploughing

Ploughing match rules will vary slightly in some countries affiliated to the World Ploughing Organisation. Details of any variations are available from the relevant country's Ploughing Association listed on page 128.

Please note, these are general rules for all classes. Detailed rules for each class are available from the Society of Ploughmen.

1. PLOUGHS

a. Tractor ploughs to be fitted with skimmers (except oat seed furrow ploughs).

b. Tailpieces are allowed (see SOP Official Rules for specifications for each class).

c. Plough bodies can be raised or lowered but cannot be raised out of the 'ploughing' position.

d. All plough bodies must be used when ploughing the finish.

e. Extraneous attachments that manipulate the furrows or scratches are not allowed.

f. In vintage hydraulic and classic ploughing, only one plough wheel allowed.

2. TIME ALLOWED

a. Tractor ploughing – 20 minutes for opening. Penalty for failure to finish opening on time – 2 points per minute or part minute.

b. Tractor ploughing – 40 minutes to judge opening.

c. Tractor ploughing – 3 hours to complete the plot.

d. Vintage and horticultural ploughing – 4 hours to complete the plot.

e. Oat seed furrow tractor ploughing – 5 hours to complete the plot.

f. Horse ploughing – 6 hours to complete the plot.

g. 15 minutes extra time for any cast-off done after the ploughing has commenced.

h. Penalty for failing to finish plot at finishing signal – 10 points per minute or part minute.

i. Maximum of one hour only allowed for breakdown.

3. OPENING

a. Conventional, vintage, horticultural and Ferguson ploughing – double opening – no land shall be left unturned.

b. Reversible ploughing – single furrow opening with right-hand bodies.

c. Classic reversible ploughing – single furrow opening made with rear right-hand body cast towards next lower numbered plot.

d. Oat seed furrow ploughing and horse ploughing – there is no opening.

e. In vintage, classic, horticultural and Ferguson ploughing on grassland only – there is no opening.

4. START

a. Conventional, vintage, classic, high-cut tractor and Ferguson ploughing – start to consist of 11 or 12 furrows.

b. Reversible ploughing – start to consist of 8 or 9 furrows.

c. Horse ploughing – start to consist of 8 furrows.

d. Horticultural ploughing – start to consist of first 6 furrows.

e. Competitors to cast towards the next highest number plot.

5. FINISH

a. Conventional, vintage, classic, high cut tractor, Ferguson and horse ploughing – finish to consist of last 8 furrows.

b. Reversible ploughing – finish to consist of last 19 or 20 furrows. Incorrect number of furrows – penalty 20 points deduction.

c. Classic reversible ploughing – finish to consist of last 11 or 12 furrows. Incorrect number of furrows – penalty 20 points deduction.

d. Horticultural ploughing – finish to consist of last 6 furrows.

e. In conventional, vintage, classic, high cut tractor, Ferguson, horticultural and horse ploughing – the last furrow to be ploughed towards competitor's own start.

6. DEPTH

Depth to be reached on completion of start and adhered to until commencing the finish.

a. Reversible and conventional tractor ploughing – minimum depth 20 cm

b. Classic vintage tractor ploughing – minimum depth 17 cm.

c. Trailing and hydraulic vintage tractor ploughing – minimum depth 15 cm.

d. Oat seed furrow and horse ploughing – minimum depth 13 cm.

e. Horticultural ploughing – minimum depth 12 cm.

7. ONE-WAY PLOUGHING – REVERSIBLE PLOUGHS

a. The competitor should make a scratch for the last 19 or 20 furrows whilst openings are being judged.

After the opening split has been judged, the competitor proceeds to:

b. Plough 8 or 9 furrows towards the lower numbered plot.

c. Start at the next highest number and complete the 'butts'.

d. Complete the last 19 or 20 furrows.

e. Alternate bodies to be used at all times. Each run to be completed before the plough is turned over.

f. The use of sighting poles at the start of the butts is not allowed.

g. When ploughing the butts, competitors cannot travel to the end of the plot to turn.

NOTE: When ploughing the joining furrow, attachments are not allowed, but it is permissible to show one or two full furrows on the first run.

8. OAT SEED FURROW PLOUGHING

Boats, smoothers, press wheels and chains may be used only in oat seed furrow ploughing.

9. VINTAGE, CLASSIC AND CLASSIC REVERSIBLE PLOUGHING

a. Plough bodies must be made for the plough at the time of manufacture.

b. Skimmers and tailpieces optional.

c. Tailpieces should not be adapted to manipulate the furrows or scratches (see rule Vintage 2e, Classic 2d, Classic Reversible 2d).

10. GENERAL

a. All ploughmen must start at the same end, as designated by the organising committee.

b. The first and second runs next to the neighbours plot will be allowed as straightening furrows and will not be judged.

c. Rolling with the tractor wheels is not allowed – penalty 20 points deduction.

d. Twin tractor wheels are not allowed.

e. Three sighting poles only (one can be placed on the headland) – assistance to set and remove is allowed. No other assistance allowed – including no assistance by signal, radio contact, mobile phone or any other method.

f. Handling, treading or shaping of scratches or furrows is not allowed. Except in oat seed furrow ploughing and horse ploughing where scratches can be handled but not removed.

g. Only one tractor and plough wheelmark allowed at the finish in tractor ploughing.

h. The use of Global Positioning Technology, laser beams, or any other electronic or computerised measuring devices are not allowed.

i. Only the competitor is allowed to clean the plot of straw, stones, etc. after the parade. Clearing of stubble is not allowed and will be penalised.

j. Any abuse or dispute by a competitor with an official of the Society of Ploughmen or a fellow competitor is not acceptable – penalty 20 points deduction.

11. FINAL DECISION

Any questions arising and not provided for in these rules shall be decided by the Executive Committee, whose decision shall be final and binding.

Judges' Scoring System

CONVENTIONAL PLOUGHS

The opening	
Well cut, uniform and straight	20
The start (11 or 12 furrows)	
Uniform and level	20
Seed bed	
Weed control and soil made available	20
Firmness	
Firm, well-packed furrows with no holes	20
Uniformity	
Clearly defined uniform furrows, and no pairing of furrows	20
Finish	
Uniform and shallow	20
Ins and outs	
Accurate and regular	20
Straightness	
Straightness of the whole plot	20
General appearance	
Workmanship and general appearance of the whole plot	20

NOTES

1. Judging for firmness, seedbed and uniformity for reversible ploughing covers the whole plot, i.e. the start furrows, butts and finish.
2. Classic reversible class plough parallel plots and use the conventional plough scoring system is used.

IN THE CASE OF A TIE

A countback system is used in all classes when two competitors have the same number of points.

Conventional Ploughing

The ploughman to win will have the highest number of points in one aspect, taken in the following order:

a) General appearance; b) Uniformity; c) Seedbed; d) Firmness; e) Finish; f) Start; g) Ins and outs; h) Opening.

Reversible Ploughing

The ploughman to win will have the highest number of points in one aspect, taken in the following order:

a) General appearance; b) Uniformity; c) Seedbed; d) Firmness; e) Finish; f) Start; g) Accuracy of final furrow; h) Joining furrows and butts; i) Ins and outs; j) Opening.

REVERSIBLE PLOUGHS

The Opening	
Well cut, uniform and straight	20
The start (8 or 9 furrows)	
Uniform and level	20
Seed bed	
Weed control and soil made available	20
Firmness	
Firm, well-packed furrows, with no holes	20
Uniformity	
Clearly defined uniform furrows, and no pairing of the furrows	20
Joining furrows and butts	
Neat and accurate with no dips or mounds	20
Finish (last 19 or 20 furrows)	
Uniform and shallow	20
Accuracy of final furrow	
Close to first furrow with no unploughed or re-ploughed land	20
Ins and outs	
Neat, accurate and regular	20
Straightness	
Straightness of the whole plot	20
General appearance	
Workmanship and general appearance of the whole plot	20

PENALTY DEDUCTIONS

Fault	Penalty
Two wheel marks	20 points
Finish the wrong way	20 points
Depth infringement	20 points per cm less than specified depth
Rolling of furrows	20 points
Failure to finish opening on time	Two points per minute or part minute
Failure to finish plot by finishing signal	Ten points per minute or part minute
Failure to report for plough-off parade	20 points
Failure to adhere to Safety Policy	Disqualification

Glossary

Bar-point share A share with an adjustable point which can be moved forward to compensate for wear. It may be secured with a bolt or be spring loaded.

Beam The main member of a plough frame running from front to back to which other parts are attached.

Boat Not part of a plough. Used in oatseed furrow ploughing to improve the appearance and shape of the turned furrows. Seamers and press wheels may also be used for this purpose.

Butts Short furrows or short work on a plot of unequal dimensions where the land runs to an angled scratch furrow or headland.

Cast off Land that is turned away from the opening towards the neighbour's plot.

Casting Ploughing away from two openings and turning left-handed on the headland.

Classic tractor classes Ploughing classes using a mounted plough hitched to a tractor in production between January 1st 1960 and before the introduction of the Q cab in 1976.

Comb Also known as Arris. The apex or top edge of the sod or furrow slice.

Connecting furrow In reversible ploughing this is the first furrow turned against the butts. Also known as the joining furrow.

Coulter A knife or disc used to make the vertical cut for the furrow slice.

Crown Also known as the top, ridge, middle, rig or cop. It consists of the first six heavy furrows on each side of the opening.

Finish Consists of the last three rounds of ploughing and the sole furrow in conventional ploughing classes.

Frog The part attached to the bottom of the plough leg to which the mouldboard, share and landside are bolted.

Furrow The space left when the sod or furrow slice has been turned over.

Gathering Ploughing around an opening, turning the furrows towards each other and turning right handed on the headland.

Heel iron Bolted to the back of the rear landside to help carry the back of the plough.

In The point at which the plough is lowered into work at the headland mark.

High cut See oatseed furrow.

Joining furrow In reversible match ploughing this is the first furrow turned towards the butts after they have been ploughed. Also known as the connecting furrow.

Landside Absorbs the thrust of the plough against the furrow wall.

Leg — The part connecting the body to the plough beam.

Mouldboard — The part of the plough which turns the cut strip of soil to form the furrow slice. Sometimes known as the breast.

Oatseed furrow — Also known as high cut or high crested work. The furrow slices are narrow and stand tightly packed together for sowing cereal seed by hand.

Opening split — The first two runs of the plough turning the furrows outwards.

Out — The point at which the plough is lifted out of work at the headland mark.

Pairing — Ploughing two unequal furrows on the same run.

Point — The removable front section of a two-piece share.

Press wheel — See Boat.

Scratch — Also known as rippling, scratch furrow or scribesod. This is a shallow furrow used to mark headlands and the butts in reversible match ploughing.

Seamer — See Boat.

Seedbed — The soil made available for subsequent cultivations to prepare a seedbed. Also known as the flesh.

Setting out — Lining up sighting poles for the first opening run.

Share — The main cutting part of the plough body. The type of share varies with the style of ploughing. Also known as the sock.

Shin — The renewable leading edge of some types of mouldboard. Sometimes called the cutter.

Skim coulter — A miniature plough body used to cut and turn the top corner from the furrow slice into the bottom of the previous furrow.

Slade — A horizontal plate, usually part of the landside, running on the furrow bottom used with horse ploughs and trailed tractor ploughs.

Sod — The strip of land or furrow slice cut out and turned over by the mouldboard to leave a furrow.

Sole furrow — The last furrow slice cut from the furrow bottom left on the previous run of the finish. Also known as a scolt or soil furrow.

Start — The first 11 or 12 furrows ploughed around an opening.

Tailpiece — A flat piece of steel fitted to extend beyond the end of the mouldboard which can be set to push down the furrow slice. Also known as a mouldboard extension.

Vintage classes — Classes for trailed and hydraulic mounted ploughs hitched to tractors in manufacture before January 1st 1960. Minimum ploughing depth 15 cm.

Wing — The rear part of a two-piece share used on some semi-digger and digger bodies which makes the horizontal cut.

About the Author and Society of Ploughmen

Brian Bell MBE

A Norfolk farmer's son, Brian played a key role in developing agricultural education in Suffolk from the 1950s onwards. For many years he was vice-principal of the Otley College of Agriculture and Horticulture having previously headed the agricultural engineering section. He established the annual 'Power in Action' demonstrations in which the latest farm machinery is put through its paces and he campaigned vigorously for improved farm safety, serving for many years on the Suffolk Farm Safety Committee. He was secretary of the Suffolk Farm Machinery Club for 60 years until retiring in 2020. In 1993 he retired from Otley College and was created a Member of the Order of the British Empire for his services to agriculture. He is past secretary and chairman of the East Anglian branch of the Institute of Agricultural Engineers. Brian's writing career began in 1963 with the publication of *Farm Machinery* in Cassell's 'Farm Books' series. In 1979 Farming Press published a new *Farm Machinery* textbook, which is now in its fifth enlarged edition, with 35,000 copies sold. Brian's involvement with videos began in 1995 when he compiled and scripted *Classic Farm Machinery Vol 1*. Brian Bell has also written on machinery past and present for several specialist magazines. He lives in Suffolk with his wife Ivy. They have three sons.

Books and DVDs by Brian Bell

Books

Farm Machinery 6th Edition
Seventy Years of Farm Machinery – Seed Time
Seventy Years of Farm Machinery – Harvest
Seventy Years of Garden Machinery
Seventy Years of Farm Tractors
Machinery for Horticulture (with Stewart Cousins)
Ransomes, Sims and Jefferies
The Tractor Ploughing Manual

Videos / DVDs

Acres of Change
Classic Combines
Classic Farm Machinery Vol. 1 1940–1970
Classic Farm Machinery Vol. 2 1970–1995
Classic Tractors
Harvest from Sickle to Satellite
Ploughs and Ploughing Techniques
Power of the Past
Reversible and Conventional Match Ploughing Skills
Steam at Strumpshaw
Thatcher's Harvest
Tracks Across the Field
Vintage Match Ploughing Skills
Vintage Garden Tractors

THE SOCIETY OF PLOUGHMEN

The Society of Ploughmen was founded in 1972 to organise the annual, British National Ploughing Championships in a different part of Great Britain each year. More than 250 ploughing societies are affiliated to it, who organise their own local ploughing matches and it has a membership drawn from ploughmen and women from all over the world. Good ploughing still lies at the heart of proper soil management and the Society's aim is to encourage and promote the art and skill of ploughing. The Society of Ploughmen also organises training and ploughing seminars for judges and for ploughmen and women who want to improve their skills and learn more about competition ploughing.

Aims of the Society of Ploughmen:

- To promote and encourage the art, skill and science of ploughing the land.
- To promote an Annual British National Ploughing Championship to be held each year in Great Britain.
- To organise training sessions for those wishing to learn more about the art, skill and science of ploughing the land.
- To co-operate with similar organisations in other countries in organising World, European and other international ploughing championships.
- To provide facilities where the premier winners at local ploughing matches who reside in Great Britain can compete for the British National Ploughing Championship and by such means foster and maintain a high standard of ploughing skills.

Contacting the Society

The Society of Ploughmen Ltd, Quarry Farm, Loversall, Doncaster, South Yorkshire DN11 9DH United Kingdom.

Website: www.ploughmen.co.uk

Phone +44 (0)1302 852469
Email info@ploughmen.co.uk

OTHER PLOUGHING ORGANISATIONS THROUGHOUT THE WORLD

Australia
National Ploughing Association of Australia –
loughfarm@iinet.net.au

Austria
Österreichische Arbeitsgemeinschaft für Landjugendfragen –
oelj@landjugend.at

Belgium
Belgisch Ploeg Comité – annemarie.claeys@telenet.be
www.belgischploegcomite.be

Canada
Canadian Plowing Organization – timbersfarm@gmail.com
www.canadaplowing.ca

Croatia
Hrvatska Udruga Za Organizaciju Natjecanja Oraca –
zalazaksunca@gmail.com

Czech Republic
Spolecnost pro Orbu Ceske Republiky – maly.a@centrum.cz
www.orba-cr.cz

Denmark
LandboUngdom – LNM@seges.dk
www.landboungdom.dk/ploejning/

Estonia
Estonian Ploughing Association –
margus.ameerikas@balticagroestonia.com
www.eestikynniselts.ee

Finland
SM-Kyntö ry – nakki.ismo@gmail.com

France
JA France Labour – jean-baptiste.moine3@orange.fr
www.francelabour.fr

Germany
Deutscher Plügerrat EV – wolf@maschinenring-ulhdh.de
www.pluegerrat.de

Hungary
Hungarian Ploughing Association – agsproginvest@gmail.com

Kenya
Kenya Ploughing Organization – aiyabeirich@gmail.com

Kosovo
Kosovo Ploughing Organization – bekimh46@gmail.com

Latvia
Latvijas Araju organizacija – boilakbaltic@gmail.com

Lithuania
Lithuanian Ploughmen Association –
alfonsas.malinauskas@lzukt.lt

Macedonia
Narodna Tehnika Na R. Makedonija
peremitrev@yahoo.com

Netherlands
Stichting Ploegvereniging Nederland – info@ploegvereniging.nl
www.ploughing.nl

New Zealand
New Zealand Ploughing Association – sjajmc@farmside.co.nz
www.nzplough.co.nz

Northern Ireland
Northern Ireland Ploughing Association –
info@niploughing.com
www.niploughing.com

Norway
Norsk Pløying – rasmusm@hesbynett.no
www.norskploying.no

Republic of Ireland
National Ploughing Association of Ireland Ltd –
annamarie@npa.ie
www.npa.ie

Russia
National Ploughing Organization – ruplow@gmail.com
www.ploughing.ru

Scotland
Ploughing Championships (Scotland) Ltd –
scotplough@btinternet.com
www.scotplough.co.uk

Slovenia
Zveza za tehnično kulturo Slovenije/Association for Technical
Culture of Slovenia – jozef.skolc@zotks.si

Sweden
Jordbrukare-Ungdomens Förbund – juf@juf.se
www.juf.se

Switzerland
chweizerische Plüger-Vereinigung –
kaethy.angst@wettpluegen.ch
www.wettpluegen.ch/

United States of America
United States Ploughing Organization – csgruber6@meltel.net
www.usapo.org

Wales
Cymdeithas Aredig Cymru – heather@rwas.co.uk
www.welshploughing.com

World Ploughing Organisation
annamarie@npa.ie
www.worldploughing.org